ROUTLEDGE LIBRARY EDITIONS:
URBANIZATION

Volume 6

URBANIZATION IN
SOCIALIST COUNTRIES

URBANIZATION IN SOCIALIST COUNTRIES

JIŘÍ MUSIL

LONDON AND NEW YORK

First published in 1981 in Great Britain by Croom Helm Ltd

This edition first published in 2018
by Routledge
2 Park Square, Milton Park, Abingdon, Oxon OX14 4RN

and by Routledge
711 Third Avenue, New York, NY 10017

Routledge is an imprint of the Taylor & Francis Group, an informa business

© 1980 M. E. Sharpe, Inc.

All rights reserved. No part of this book may be reprinted or reproduced or utilised in any form or by any electronic, mechanical, or other means, now known or hereafter invented, including photocopying and recording, or in any information storage or retrieval system, without permission in writing from the publishers.

Trademark notice: Product or corporate names may be trademarks or registered trademarks, and are used only for identification and explanation without intent to infringe.

British Library Cataloguing in Publication Data
A catalogue record for this book is available from the British Library

ISBN: 978-0-8153-8014-6 (Set)
ISBN: 978-1-351-21390-5 (Set) (ebk)
ISBN: 978-0-8153-7928-7 (Volume 6) (hbk)
ISBN: 978-0-8153-7935-5 (Volume 6) (pbk)
ISBN: 978-1-351-21614-2 (Volume 6) (ebk)

Publisher's Note
The publisher has gone to great lengths to ensure the quality of this reprint but points out that some imperfections in the original copies may be apparent.

Disclaimer
The publisher has made every effort to trace copyright holders and would welcome correspondence from those they have been unable to trace.

URBANIZATION IN SOCIALIST COUNTRIES

Jiří Musil

URBANIZATION IN SOCIALIST COUNTRIES

With a foreword by Simona Ganassi Agger

M.E. Sharpe, Inc.
WHITE PLAINS, NEW YORK

Croom Helm
LONDON

Copyright © 1980 by M. E. Sharpe, Inc.
901 North Broadway, White Plains, N.Y. 10603

All rights reserved. No part of this book may be reproduced in any form without written permission from the publisher.

First published in Great Britain 1981
by Croom Helm Ltd
2-10 St John's Road, London SW11

This book is a revised and expanded version of Chapter 7 of *Urbanizace v socialistických zemích*, Prague, Nakladatelství Svoboda, 1977.

Published simultaneously as *International Journal of Sociology*, vol. x, no. 2-3, 1980 (Summer-Fall).

M. E. Sharpe, Inc. ISBN: 0-87332-180-4
Croom Helm ISBN: 0-7099-1719-8

Printed in the United States of America

TABLE OF CONTENTS

List of Tables
vi

Foreword by Simona Ganassi Agger
vii

Introduction
3

Czechoslovakia
17

The Soviet Union
45

Poland
75

The German Democratic Republic
100

Hungary
113

Romania and Bulgaria
127

Yugoslavia
141

Conclusions on Settlement Strategies
148

Notes
159

Selected Bibliography
167

Index of Names
179

Index of Subjects
181

About the Author
185

List of Tables

1. Development of Urbanization in CSR and SSR — 18
2. Population Employed in Industry in CSR — 21
3. Concentration of Population by Settlement Groups in CSR — 36
4. Structure of Settlement by Size in CSR — 40
5. Differences among Urbanization Variants for Slovakia — 43
6. Urban and Rural Population of the USSR — 46
7. Selected Examples of Rapid Growth of Soviet Cities — 47
8. Growth of Cities in the USSR — 47
9. Development of Ten Largest Urban Agglomerations in the USSR — 50
10. Percentage of Population by Settlement Categories in the USSR — 55
11. Growth of Towns and Urban Population in the USSR — 55
12. Reasons for Origin of New Towns in the USSR — 65
13. Urban and Rural Population in Poland — 77
14. Proportions of Polish Population in Rural Communities, Towns, and Urban Settlements — 80
15. Urban Agglomerations in Poland — 90
16. Hypothetical Changes in Population Distribution and Urbanization Areas in Poland — 91
17. Forecast for Industrial-Urban Agglomerations in Poland — 95
18. Population by Sizes of Communities in the GDR — 103
19. Population of Budapest as Percentage of Total Hungarian Population — 116
20. Population of Budapest and Its Agglomeration — 117
21. Population by Sizes of Communities, Hungary — 120
22. Rural and Urban Population of Romania — 128
23. Proportion of Population of Towns by Size Groups and Its Growth in Romania — 131
24. Population by Sizes of Communities in Bulgaria — 135
25. Population by Sizes of Communities in Yugoslavia — 142
26. Projection of Urban Population Trends in the Socialist Countries — 150

FOREWORD
By Simona Ganassi Agger

For the Western urban planner, urban sociologist, or "urbanist," as the European might say, this book is essential reading for understanding something of the complex situation in urban policy development during the past twenty or so years in the socialist countries of Europe. It is exceptional in its breadth of coverage:—only Albania receives no consideration in these pages.

Although the author is limited in how much analysis he can do while presenting descriptive materials on the urban policy situation in eight countries, he has made a remarkable synthesis of an incredible amount of materials on each country. Moreover, his analytic framework does permit him to reveal to the reader some very fundamental issues these countries have faced and continue to face. In the process the Western reader begins to understand, as a kind of bonus, something of the differences and similarities between the professions of urban planning in the United States or typical Western European countries and those in Eastern European countries as well as the Soviet Union.

Perhaps the most striking, but not surprising, difference is that socialist urbanization is regarded as a planned and managed process. Not only the economies but also the settlement systems are planned. This fits the basic philosophy of the social and political structures of these countries. It is, however, surprising how similar are some kinds of conceptual and technical problems, at least, that the urbanist in both the West and East faces in trying to indicate to policymakers what to do and how to do it.

The author's own specialization as an urban sociologist may have sensitized him to the difficulties that planners everywhere seem to have in coping with social and economic goals simultaneously. In fact there emerges from his description and analysis a sense that the economic efficiency of the settlement system becomes a conditio sine qua non for achieving social goals.

Dr. Musil makes it clear that while such concepts as "rational" and "optimal" are as common in the socialist urbanist's vocabulary as they are in the lexicon of the modern American urban systems analyst, they still need clarification and must be seen in the concrete social and political context. This means that various interests and perspectives may define them very differently. Possibly because the American urbanist is further removed from the play of those who make and remake the urban landscape and settlement, she or he is not as caught up in the theoretical and conceptual controversies of European urbanists, whose varied approaches become more of a vehicle for opposing policy views and interests. A viewpoint that opposes a currently authoritative school of urban thought is likely to bring desired notoriety to a professional or academic urbanist in a country such as Italy or France, thus stimulating much intellectual conflict. In other countries such doctrinal differences are perhaps best muted. But intellectual diversity seems endemic everywhere.

As Dr. Musil shows, one senses a shift in general approaches to the urbanization process in the socialist countries. He mentions in his concluding chapter, as well as principles that have remained unchanged, those shifts in urbanization strategies of fundamental importance that occurred in the space of only ten to fifteen years.

I expected to find that Yugoslavia was a country of constant

Foreword

policy experimentation, including with urban policies. And I anticipated that the Eastern European countries, including the Soviet Union itself, had pursued a more steady kind of line in the urbanization strategy. I was surprised, as I think many readers will be, that my latter expectation was wrong.

Dr. Musil's chapter on the Soviet Union is particularly valuable. Not only does he give the reader a clear picture of what kind of urban policy developments have occurred there since the Revolution, he is appreciative of the problems urban analysts face there — as everywhere — because of the lack of linkages between specialized sectors and physical planning as well as among the other divisions of labor. His sense that the Soviet theorists and policymakers are now embarked on a linking venture through systems analysis that models and tries to manage such complex things as a national or republic settlement pattern is illuminating. Equally illuminating is his discussion of present Soviet principles, which, while sophisticated both in terms of complex system modeling and simulation and in recognizing the many real problems in any such approach, do not seem to question sufficiently the deeply rooted principle of specialization based on divisions of labor in space as well as in the organization of the economy.

Dr. Musil informs us in a precise, penetrating fashion how socialist urbanists generally had to and still have to deal with the central socialist principle of equality. It has been a conceptual and intellectual problem of the first magnitude to translate equality into human settlement patterns. Partly because the latter are congeries of dimensions and factors, partly because there is so much still unknown about how land-use patterns affect people, or, indeed, if such a traditional urban planning problematic makes sense at all, and partly because the concept of equality is so packed with meanings, one is faced with numerous modes of translating it into spatial densities and uniformities over varying time frames (immediate present, short-term or longer-term futures, etc.).

Simona Ganassi Agger

As the author indicates, an earlier notion emphasizing uniformity in terms of equal densities of population is no longer an operative assumption in any of the countries surveyed. There is now a commitment to a uniformity of social qualities or to reducing inequalities as much as possible. How the latter relates to various patterns, policies, and strategies of human settlements in these countries is a fascinating subject just touched on here.

How social equality and other criteria of urban policy fit together differs according to specific and quite variable conditions from socialist country to socialist country. Some countries with great regional disparities in economic conditions, such as Poland or Yugoslavia, obviously must have a different strategy than a Czechoslovakia or a German Democratic Republic, which have relatively fewer. Yet, as Dr. Musil notes, the aforementioned evolution in approach occurred, so that in recent years a growing emphasis has been placed on the economic significance of settlement structures, with stress on concentration of economic and social activities but no longer, or as much, on population or housing concentration.

The multitude of economic-industrial development policies as well as urban settlement strategies in these eight countries provides a wealth of experience for other countries that seem almost fated in the future to have a national rather than a traditionally decentralized urban policy approach. National urban policy may not escape a piecemeal pragmatism as urbanists and others grope for the most appropriate measures in a sea of such measures, as we can see from this book. Almost paradoxically, tolerance and flexibility about the best urban forms increase just as the importance of the settlement fabric to national macroeconomic goals increases. But with the energy crisis and the crunch coming on various other natural resources, as well as growing concern with environmental qualities, one does not have to be a particularly acute prophet to sense the coming of increasingly national commands and controls even in countries that may temporarily move more toward free enterprise than state planning.

Foreword

Lessons learned in the socialist world may be of some benefit to other countries, especially in avoiding urbanization strategies that seem superficially promising but which have been revealed elsewhere as having hidden conceptual or practical problems. It would profit us all if Dr. Musil would soon write another book on the same subject, with the same kind of clarity and perceptive analysis, but at greater length and depth and concentrating on the lessons that are being or could be learned from these midtwentieth century policies in testing and then changing approaches for which theory was nonexistent or conflicted with practice.

May 1980
Istituto Universitario
Architettura Venezia and
Eastern Kentucky University

URBANIZATION IN SOCIALIST COUNTRIES

INTRODUCTION

Urban and regional studies have undoubtedly garnered increased attention recently in the socialist countries. Characteristic of the past ten to fifteen years have been the production of many prognostic studies and the emergence of complex developmental programs, sometimes described as urbanization strategies, general projects for future settlement, or national settlement policies. We shall focus here on the concept of urbanization strategy, which entails the formulation of general principles and methods for developing a national settlement system. In most cases it also includes selecting main and secondary growth centers, studying data on the anticipated concentration of urban population, and generating principles for handling macroregional problems in particular countries.

A vast amount of empirical, analytical, prognostic, and theoretical material from many fields has had to be examined in order to formulate such strategies. The published matter on the urbanization process ultimately yielded an almost unmanageable abundance of facts, interrelations, and problems — the assembled findings had to be sifted, compared, and evaluated. Attempts to produce such a survey have been largely confined to individual socialist countries; relatively few publications have dealt with the socialist countries of Europe collectively, although the need for such a wider view was already evident in the 1970s. The published work includes, for instance, K. Mihailovič's study of regional development in Eastern Europe; Iu. L. Pivovarov's work on contemporary urbanization, which has chapters on urbanization strategies in the socialist countries; and a collection edited by A. Kukliński on regional planning in European countries, which contains contributions on some of the socialist countries.

The present study intends to survey, systematize, and consider the evolution of urbanization strategies through the 1960s and 1970s in individual socialist countries. It is a revised version of the final chapter of my book <u>Urbanizace v socialistických zemích</u> [Urbanization in the Socialist Countries], published in Prague in 1977, which dealt with the social, economic, and planning aspects of this subject. In contrast to the above-mentioned synthetic works on regional development, the focus here is on the system of settlement and on such problems as its regulation, as well as analysis of the goals, instruments, and techniques used in planning the urbanization process in different socialist countries. Regional conditions and problems naturally constitute the "environment" that has to be understood when dealing with urbanization, but they have not in themselves been at the center of interest in this survey.

My aim is to provide as much information as possible that can throw light on the basic premises underlying the formulation of urbanization concepts, revealing their main features and indicating their main lines of development. Since the authors of the strategies were faced with problems differing according to the conditions in their respective countries, I have introduced each section with a brief description of the regional and settlement problems involved.

All the socialist countries of Eastern Europe are dealt with, with the exception of Albania, where the inaccessibility of the literature and the language barrier have made inclusion impossible. Although I have tried to give equal weight to the material from each country, it is probable that some countries have received more, some less attention. For the most part these discrepancies can be accounted for by the fact that in some socialist countries the subject has been more fully studied

Introduction

and more has been published; some of the imbalances — especially with regard to the countries of southeastern Europe — are due to my inadequate knowledge of the languages and the smaller number of available sources. Finally, my work was influenced by the fact that I am a sociologist; although I have tried to give due weight to economic and planning aspects, I have probably not avoided interpretations made to a considerable degree from a sociological standpoint.

Since many of the facts and situations mentioned in the survey would be difficult to understand without reference to the broader examinations of urbanization problems that have been published in the socialist countries, this introduction also includes some brief information about the main subjects covered by urbanization studies over the past two decades. Knowledge of these contexts is in many cases the key to understanding the particular strategies and their development. The subjects include: (1) a new concept and definition of urbanization; (2) the relation of urbanization to industrialization and modernization; (3) urbanization and the scientific-technological revolution; (4) urbanization as cultural process and the diffusion of the urban way of life; (5) processes of concentration and how they are changing; (6) the economic effects and social consequences of concentration; (7) integration of the economic and social goals in urbanization strategies; (8) linking central place theory with the growth poles.

Toward the close of the 1960s and the beginning of the 1970s there was a revival of interest in clarifying the content of the term "urbanization"; several studies were published with a view to redefining this process. Definition was not merely a theoretical matter; it also bore on ideas about how to regulate the process. The result of that discussion was a move away from so-called narrow definitions, which saw urbanization purely as a process of population concentration, as the growth of cities, or perhaps as a change in the material and spatial arrangement of human settlement. New definitions interpreted urbanization as a complex and universal societal process that is intimately linked with the development of productive forces and with forms of social communication, transforming the sociospatial organization of society as a whole.

Although urbanization is today a universal process, its specific forms are differentiated by the economic, social, and political conditions prevailing in the societies where it takes place. In the socialist countries it is seen as a planned and directed process; and especially in the publications of the seventies, it is termed "managed urbanization." Managed urbanization presupposes the working out of a general strategy embracing both goal-setting and the instruments to be used. Research in the socialist countries is concerned, then, not only with analyzing the process itself but also with formulating normative principles for its advance. They provide the basis for planning.

The management of urbanization depends to a large extent on understanding how the urbanization process relates to industrialization and modernization. Thus the urbanization-industrialization connection assumes key importance. Attention had been directed to the fact that these two macrosocial processes may be more or less closely linked and to how the relationship between them changes during different phases of the urbanization process. Problems for discussion therefore included the question of the extent to which the industrialization process determines urbanization, and how that determining influence is gradually moderated. In this connection intensive investigation of the interaction between the scientific-technological revolution and urbanization had already started in the early seventies — in fact, the most recent urbanization strategies originated in the demands of those days for "intensification" of the economy, for the optimal location of scientific institutions, and in the increasing influence of information processes on shaping the settlement system. Important from the sociological standpoint are the Hungarian, Polish, and Czechoslovak studies on the so-called lag of urbanization behind industrialization, as well as those on how urbanization processes diverged in being accompanied by industrialization or, on the contrary, "preceding" it. Empirical studies in this field have

Introduction

produced a typology that no longer views urbanization as a single, undifferentiated process. In connection with the rapid industrialization of the socialist countries, particularly of those which after World War II were among the less developed, and with predominantly agrarian economies, the process of migration to urban communities and the many social problems caused by major shifts of population to urban areas have been investigated. One product of research in this field has been the concept of the ruralization of cities, which developed parallel to that of the urbanization of the countryside and the "peasanting" of workers. More rapid increase in industrial job opportunities, accompanied by slower concentration of population in towns, has led to the growth of a category of workers who have jobs in the towns but continue to live in the country. The social and cultural consequences of this combination have been a subject of intensive study, particularly in Poland, with attention directed both to this specific category in the working class and to the communities where these peasant-workers live.

Compared to the interest aroused by study of the relation between urbanization and industrialization, the interaction between urbanization and modernization has received only modest attention. Insofar as the concept of modernization has been used in this regard, it has been in Polish and Romanian work concerned with diffusion of some elements of urban life style into the countryside. In countries where the difference between urban and rural communities is not great, such as Czechoslovakia and the German Democratic Republic, modernization processes have not attracted much attention, or they are studied in other contexts — most frequently that of the social and cultural changes accompanying urbanization in society as a whole.

In countries which have undergone or are now in the process of rapid industrialization, and which have had strong traditions of rural culture, research on the social and cultural transformation of the countryside is well developed. It is strongest in Poland, promoting a very effective combination of urban and

Urbanization in Socialist Countries

rural sociology. Similar studies have been published in the Soviet Union, Czechoslovakia, Romania, and Yugoslavia. The most frequent themes are changes in the social structure of formerly purely rural areas, changes in the structures and functions of social groups, especially families, and changes in the traditional forms of social interactions and cultural patterns. The methods used are very diverse: for instance, monographic analyses of social change in individual towns, research on the changes occurring in families where one or more members commute from country homes to work in cities, or observations of social and cultural changes in rural communities situated close to cities.

The growing interest in the social and cultural aspects of the urbanization process that is quite evident in current sociological work in the socialist countries does not mean that efforts to gain a better understanding of the socioeconomic implications of urbanization have been pushed into the background. Along with theoretical studies on such traditional themes as the location of productive, particularly industrial, forces or the influence on the urbanization process of technological advance, the division of labor, or concentration, specialization, and cooperation in industry, one can observe a shift in emphasis to some topical subjects and to problems that are directly or indirectly connected with the formulation of regional policy and urbanization strategy. The spectrum is quite broad, and I can mention only some of the most important points.

Most deserving of attention are studies concerning the macroeconomic effects of the ways in which settlement networks and their elements are arranged. They relate to varying concentrations of production, infrastructure, and population, the effects of proximity among economic, political, cultural, and scientific institutions, and the effects of the spatial arrangements of settlements themselves. A decisive influence in shaping views concerning the desirable trend of settlement systems has been exerted by studies on agglomeration effects in industry, optimal sizes for different infrastructure components, costs of construction and management for urban com-

Introduction

munities of different sizes, and research on the economic effects of industrial agglomerations and of investments in them. According to the conditions in a given country, the main emphasis has always been placed on some specific problem. For instance, in countries with large industrial agglomerations (the USSR, GDR, Poland), much attention is devoted to their macroeconomic effects; in countries with dense networks of medium-sized and small cities, the emphasis is on optimal distribution of industry and infrastructures within the existing settlements, while agglomeration effects in production and optimal sizes for different types of social infrastructure are also examined. Where there is extensive construction of new towns, e.g., the Soviet Union, research on construction and management costs for urban communities of varying sizes absorbs considerable attention.

Changes in agriculture have had their influence on the urbanization processes. Ongoing mechanization, the growing size of production units, and increasing cooperation among them, together with the organizational changes induced by large-scale farming, have made their impact not only on the transformation of individual rural communities but also on the whole system of rural settlement, including small towns. Rapid depopulation of the smallest communities, with stores and services concentrated in selected centers and small towns, the planned establishment of jointly administered communities or "community associations," the emergence of a kind of supercommunity integrated under a single administrative center (the Polish gmin) — these are all changes radically affecting the "bottom" of the settlement system. This development has stimulated much work on group-system models for rural settlement. Many studies have also been published about the possibility of reviving the small towns, most of which have stagnated. Their situation is gradually improving where they are becoming organizational centers for large-scale farming or for specialized agricultural services, and where, close to cities or incorporated into agglomerations, they are developing a residential function.

Urbanization in Socialist Countries

Increasing attention has been devoted in the period under consideration to the influence of the tertiary sector on shaping the settlement system. Questions investigated include optimal sizes for facilities within the social infrastructure, e.g., hospitals, schools, libraries, etc., and the rational concentration or deconcentration of social infrastructure components; in general, increased attention has been paid to how development of the tertiary sector affects the growth of cities. New themes include the more rational location of high schools, universities, and scientific institutions.

The discussions on concentration have begun to differentiate between the concentration of productive and nonproductive activities, population, and information and decision-making processes; the whole problem is now beginning to be seen more fully than before as an evolutionary process. This has also influenced the construction and content of urbanization prognoses and strategies. Population concentration in towns, long regarded as an essential feature of urbanization, will in the future, according to work by Czechoslovak geographers and sociologists, be replaced by the concentration of certain progressive activities (e.g., some of the more modern industrial sectors, services, scientific institutes) and later by the concentration of information and decision-making processes. For this reason some authors already define contemporary urbanization as the concentration of functions (activities) in relatively few locations and centers where these activities are most productive. Concentration is to some extent accompanied by deconcentratration of some activities and their universalization, which is accentuated in the socialist countries by the comprehensive systems of education, health care, social welfare, etc. (Hampl 1971).

Findings from studies of concentration and of growth in "contrast" (Pivovarov 1976), as shown by the distribution of population over the territories of different countries, have influenced the formulation of principles for settlement development during the period we are considering. Urbanization strategies that in the sixties in some countries, e.g., Hungary

Introduction

and Czechoslovakia, were still aimed at evenly distributing a fairly large number of urban centers over the country gradually began to stress the need to intensify and rationalize the settlement system, i.e., among other things, to concentrate productive and nonproductive investment in selected centers or agglomerations fewer in number than envisaged by the earlier proposals. The endeavor to develop weaker regions as well while placing greater emphasis on concentration then led to a synthesis called, e.g., in the Polish concept of planning, "polycentric concentration" (Grabowiecki 1974).

A new feature in the economic research on urbanization has been a growing tendency, especially during the seventies, to establish links with the sociological and ecological approaches. Some recent examples are demonstrative.

Studies of the economic effects produced by urbanization strategy variants are being increasingly linked with simultaneous investigations of the social and ecological consequences of these variants. This move toward a more complex and synthesizing view of urbanization is among the notable current trends. It is evidenced, for instance, by the fact that in several of the socialist countries, notably in the USSR and Czechoslovakia, mathematical methods have been elaborated for complex evaluation of the effects resulting from different urbanization policies and from varying degrees of population concentration. Quantification of the effects was preceded by qualitative analysis of the advantages and disadvantages of varying degrees of concentration (Líkař and Musil 1975), with reference to the relations between settlement sizes and labor productivity in different sectors of industry, the costs of the technical infrastructure, loss of time in commuting, the health of the population, criminality, air pollution, and so on.

Linking the economic and sociological approaches to urbanization analysis also stems from the growing emphasis placed on intensive forms of economic development. Such forms are shaped, among other things, by what has been termed the psychosocial infrastructure (Ziółkowski 1967), understood as the social conditions required if a city is to attract external sources of growth; they include, for instance, regional consciousness, openness to information, an inclination to innovate, organizational ability, etc. A further incentive to integrate economic and sociological perceptions of urbanization is the fact that intensification of economic development depends on advances in the scientific and technological revolution, and that urbanization is among the processes providing favorable conditions for it. Soviet sociologists and economists (Akhiezer 1968; Akhiezer, Kogan, and Ianitskii 1969; Ianitskii 1972) have pointed to this fact, emphasizing that urban communities, especially big cities, offer the best environment for the intensification of economic processes thanks to their concentration of activities, information, and varied job opportunities, as well as to their multiplicity of personalities — the concentration of "psychologically flexible people" capable of rapidly accepting innovations.

To enable the reader to understand the problems confronting authors of urbanization strategies in socialist countries, it has been essential to map the main subject areas that have been examined in analytical and theoretical studies of urbanization over the past fifteen years or so. These subjects recur in modified form also in various normative strategies for urbanization. Before giving a more detailed account of these strategies, it will be useful to note the issues that had to be tackled in almost all the countries.

The central issue common to all the countries is how to integrate the social and economic goals of the normative strategies. The search for the optimal synthesis concerns both regional policy and urbanization strategy, especially in efforts to regulate the concentration processes. Regional development should be aimed at removing the differences between regions while at the same time yielding the best results in the macroeconomic sense. Some theories maintain that in many cases these results are obtained by investing in already developed regions with good infrastructures that have populations adapted to industrial work and equipped with advanced skills. Good re-

Introduction

sults from such investments then create resources for attaining the social goals in the areas which, for various reasons, lack the conditions for intensive development. Against this it is argued that such problems occur especially in countries with low population and urban settlement densities, and then only during certain phases of industrialization. Moreover, it is probable that this interpretation derives from a short-term view in which quick economic results are the main concern. A long-term view can lead to different conclusions. What is more, the emphasis is on industrial achievement, while the role of other economic sectors, especially modern agriculture and its associated industries in the regional process, is underestimated. What is seen by some writers as a dilemma is therefore regarded by others as a problem that can be solved, and they point to the successful elimination in socialist countries of major macroeconomic imbalances, e.g., between Slovakia and the Czech parts of Czechoslovakia. However, it must be admitted that the existence of quite considerable regional differences in Poland and Yugoslavia, for instance, shows that the problem remains a serious one.

The same question arises in another guise in connection with the concentration of activities and population in urban areas, especially large cities. Various mechanisms, mainly economic in nature, maintain the concentration of population in large towns and industrial agglomerations and conurbations. The benefits from this process are not purely economic, as the simplistic view has it; they are also social. Nevertheless there are negative social and ecological consequences as well, and the measures to prevent or eliminate them can, when the city or agglomeration is of a certain size and under certain conditions, minimize or practically cancel the economic benefits. On the other hand, there are those who maintain that the economic benefits derived from urban growth and concentration provide the resources necessary to carry out social programs and to improve the quality of the urban environment. This matter is still being debated, but it seems that the emphasis on intensifying economic development favors the idea of "rational" concentration. By this is meant, in effect, not putting a stop to growth in cities and big industrial agglomerations and allowing further growth of cities with fewer than 500,000 inhabitants as well as of smaller and medium-sized cities.

The search for variants of managed urbanization that could maximally integrate social and economic goals has forged a link between the two main theories on the development of settlement systems that prevailed during the sixties and seventies in the socialist countries. The first theory could be described as a normative version of the central place theory, its purpose being to create a fairly dense network of centers where the facilities of the social infrastructure would be concentrated. The accessibility of these centers should be such that the same basic components of services, stores, schools, health, and other facilities would be available to practically the entire population. The second theory, which can be described as a strategy for the development of growth in centers and agglomerations, is a specific form of the polarization theory. Its aim is to achieve, by planning a system of settlements, the maximum economic effect from production while simultaneously improving social and ecological conditions in the areas of higher concentration of economic activities and population. With some simplification of it can be said that the first theory gives greater weight to consumption, the second to production. In reality the two approaches have often merged, and the search for the most suitable combination of them is still the subject of discussion and research.

It was necessary in this context to clarify the general relationship between "production" and "settlement system." In some strategies, especially in the USSR, Poland, and the German Democratic Republic, the role of settlement is seen to be active, it being a relatively independent factor in creating conditions for more efficient production. Others view the function of the settlement system and the influence of changes in it as passive factors which, they insist, should be "in harmony" with the needs of production.

Views on the relationship between the social and the economic aspects of development have not been without influence

Introduction

on ideas concerning the number of centers to be developed and to be the focal points for productive and nonproductive investment. Whether more or fewer settlements are included in the category of these centers not only affects the economy of the settlement system but also has social consequences for, for instance, the volume of commuting over long distances, the extent of housing construction, the numbers of peasant-workers, and so on. Therefore a subject of much discussion among economists, sociologists, and planners is what direct and indirect results may follow from urbanization strategies differing as to the number of settlements selected for planned development and how, in the light of knowledge about these consequences, to choose the most suitable alternatives.

Finally, but not least important, I must mention a matter that both in theory and in current practice is the subject of a lively exchange of views. It concerns the relationship between regional development theories and long-range concepts of urbanization, which are not always coordinated (Mihailović 1973), as well as the relationship between sector planning and physical planning. The latter problem in particular has attracted increasing attention because underestimating either physical or regional aspects can detract from planning efficiency. Demands on transportation may be greater, the infrastructure may be made less efficient, and in the social sphere there may be created disequilibrium in the labor market, unforeseen demands on social services, housing, etc. Therefore ways are presently being sought to more effectively integrate different planning procedures. One of the ways to link sectoral economic planning with regional and physical planning is evident in the social planning of territorial units, a method being developed in the USSR, Czechoslovakia, and Poland. This has stimulated a reorientation of research and of the social science disciplines. For instance, it has been possible in recent years to observe a growth of interest in regional sociology going beyond the bounds of the traditional urban sociology, and a growing number of theoretical studies have been devoted to goal-setting problems, to goal integration, and to models of settlement systems and sophisticated methods and techniques for planning large territorial units. The urbanization strategies that will be elaborated in the 1980s will undoubtedly be enriched by the results of this work.

1

CZECHOSLOVAKIA

Taking the country as a whole, Czechoslovakia is today at a medium level of urbanization. In 1976, 55.5% of the population of the Czechoslovak Socialist Republic (ČSSR) lived in urban communities, while in the Czech Socialist Republic (ČSR) the urbanization level is higher (according to the 1970 figures from the Federal Office of Statistics, the urban population amounted to 62%, making it among the most highly urbanized areas in the socialist world).[1] Slovakia (SSR), which was the most backward, agrarian part of the prewar republic, where industrialization did not start until the period of socialist construction, is rather less urbanized, although there has been more rapid growth of towns in recent decades than in the Czech lands of Bohemia and Moravia. In 1970, 41.4% of the Slovak population lived in urban communities. The rate of change recorded by Slovakia in the period 1930-70 is evident from Table 1, which shows the types of urbanization. Rapid industrialization and urban growth in Slovakia have narrowed the gap between the two parts of the republic with regard to the nature of settlement, the differences now being smaller than between some of the Yugoslav republics or the macroareas in Poland. There can be no doubt, however, that historical, demographic, and geographical factors have caused the settlement network in Slovakia to develop along lines differing from those in Bohemia and Moravia. The settlement system in these Czech lands was of the mid-European and western European type, with numerous small towns founded mainly in

Urbanization in Socialist Countries

Table 1

Development of Urbanization in ČSR and SSR

Urbanization type	Republic	1921	1930	1950	1961	1970
Basic	ČSR	29.6	32.9	41.2	46.2	52.4
	SSR	18.5	22.7	26.2	29.7	36.9
Medium	ČSR	15.8	19.5	28.2	30.5	35.5
	SSR	4.9	9.2	11.5	13.6	21.3
City	ČSR	10.4	12.7	17.2	17.7	18.9
	SSR	0.0	3.7	5.6	5.8	9.9

Source: V. Srb, Demografická příručka; Sčítání lidu, domů a bytů ČSR 1970 (census figures).

the Middle Ages and with dispersed agricultural settlement of feudal origin; whereas in Slovakia, as in the greater part of Poland, settlement was of an intermediate type, that had some eastern European features, for instance, larger rural communities and lower density.

Among the typical, historically determined features of the settlement system as it has evolved under the natural and social conditions peculiar to Czechoslovakia is a strong representation of small and medium-sized towns. This was accentuated in the early days of industrialization by the location of industries, the results of which are still evident, especially in the Czech parts of the country. In analyzing the distribution of Czechoslovakia's industry, Jaroslav Mareš writes that "typical for our industry was dispersal not only into numerous factories but also into many industrial centers. Bauer found that in 1930 factory industry existed in one third of all the administrative communities in our country. Of 3,927 industrial communities, however, over three quarters had no more than 2,000 inhabitants. The share of these industrialized villages and townships in the total horsepower in industry was at that time around 22%. That is to say, small firms predominated."[2]

In addition to being scattered, industry was also quite unevenly distributed over the country. First there was the macrodisproportion between Slovakia and the Czech lands, typified

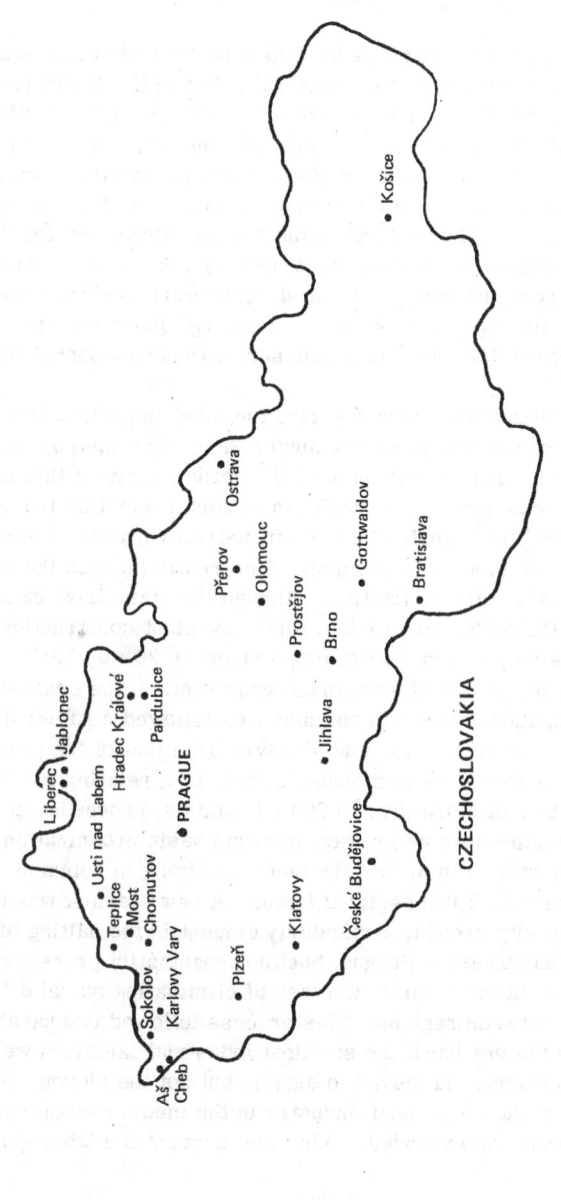

for instance by the fact that in 1930 Slovakia's share in total industrial employment was about 12%. But in the Czech parts, too, there were regional differences. Over four fifths of the industrial work force was concentrated in the 1930s in just under one third of the state territory, with almost all the industrial centers lying, with a few exceptions, north of a line running from Aš to Ostrava and in the Prague, Brno, and Plzeň agglomerations. Nevertheless, thanks to a dense communications network, relatively advanced agriculture, and the scattering of small towns that served the farming hinterland, these differences did not amount to any serious macroregional imbalance.

In the immediate postwar period the most important factor affecting settlement was the transfer of the German population and the resultant resettlement of the border regions; this involved an extensive migration of the Czech population from the interior and the transfer of some industrial capacity to Slovakia, where labor was available. This really marked the start of Slovakia's industrialization, although the main drive came later, in the fifties and sixties, when new plant construction raised its share in industrial employment to 20% by 1960.

During the period of socialist reconstruction, the concentration of population in urban communities followed a rather different course in Bohemia and Moravia from that of the prewar period. So-called medium urbanization, i.e., relating to urban communities of more than 20,000 inhabitants, proceeded at about the same rate as prewar, whereas basic urbanization was somewhat more rapid, i.e., the numbers living in communities with fewer than 5,000 declined faster. A new element was the slowing of city growth, particularly evident in the halting of population increase in Prague. Such an urbanization process corresponded, in the main, to the task of eliminating social differences between regions. This process followed two parallel lines: on the one hand, the smallest industrial factories were closed, merged, and moved to bigger, but not the biggest, locations; on the other hand, industry in the medium-sized and small towns was expanded, taking into account the labor-power

Czechoslovakia

Table 2

Population Employed in Industry in ČSR,
1930-70

Community groups by size	Percentage of population in industry and craft production				Growth indexes 1950 = 100	
	1930	1950	1961	1970	1961	1970
500-1,000	31.6	25.5	28.0	27.6	110	108
1,000-2,000	37.5	28.0	31.8	30.8	114	110
2,000-5,000	43.3	33.4	37.0	35.5	111	106
5,000-10,000	44.7	37.8	43.4	41.8	115	111
10,000-20,000	43.1	33.5	39.8	38.9	119	116
20,000-50,000	45.3	39.0	43.0	37.2	110	95
50,000-100,000	—	35.2	38.1	41.0	108	116
100,000 and over	40.3	32.5	35.0	28.2	108	87
Total	36.1	29.2	33.8	32.4	116	111

Source: V. Srb, Demografická příručka; Sčítání lidu, domů a bytů ČSR 1970 (census figures).

balance in the regions. Mareš states that the average size of industrial factories had increased 2.7 times by 1960 compared with 1930.

This strategy caused the biggest increase in the industrial work force in the ČSR to be recorded by towns of 5,000 to 20,000 inhabitants, while the smaller industrial centers and small towns also showed an average industrialization rate[3] (see Table 2). Since the industrial centers were fairly accessible, as the agricultural work force declined and a sizable non-farming population emerged, commuting to town became widespread. A reflection of this can be found in the figures for 1961 and 1970, when 34-35% of the economically active population in the ČSR traveled daily from their homes to work in another community. Although comparable data are not available for other socialist countries, it is probable that Czechoslovakia and the GDR rank among the countries with the highest levels of commutation.

Urbanization in Socialist Countries

The special features of the settlement system in the Czech and Slovak Socialist Republics, which in the Czech part include in particular a large proportion of small and medium-sized towns and a dense network of urban settlement and industrial agglomerations (smaller, however, than in the neighboring GDR and Poland), are reflected in the proposals for future development.

National Settlement Strategies in Czechoslovakia

The ČSSR ranks with Hungary and Poland among the countries where intensive work on national settlement strategies had already started at the beginning of the 1960s. The impetus came partly from the experience of the preceding years and partly from the growing need to make maximum use of the potential to be found in rational development of the country's territory. A national settlement strategy was also required because economic planning (in its regional aspects) had to be more effectively linked to territorial planning.

Another factor favoring broadly conceived national projects was the tradition of the Czechoslovak architectural avant-garde, which even in prewar days had formulated a number of ideas about socialist reconstruction of settlement — for instance, zonal development of urban settlement, building of socialist towns based on collective blocks of flats, planning the residential landscape, and so on. These ideas were reflected in various proposals that have not lost their validity even today. Just a few examples: E. Hruška's idea for developing Prague as a regional city; J. Voženílek's plan for developing the Gottwaldov agglomeration as a form of settlement belt; or Kumpošt's macroregional concept of an urbanized belt along the Morava River.

The sophisticated nature of many suggestions that emerged soon after the end of World War II reflected the "maturity" and also the complexity of the settlement structures, especially in

Czechoslovakia

Bohemia and Moravia. True, apart from the macroregional disproportion between Slovakia and the Czech lands, Czechoslovakia has not suffered any major imbalances or problems of the kind experienced by Hungary, Austria, and Denmark, which stem from the overgrowth of their capital cities; or by Britain, the Federal Republic of Germany, the German Democratic Republic, and Poland, with millions of people concentrated in industrial agglomerations; or at the other extreme, the excessively thinly settled northern regions of the USSR, Finland, and Sweden. Nevertheless it would be a mistake to overlook the complexity of the settlement system in the ČSSR when socialist construction and reconstruction began.

The building of socialist society started with Bohemia and Moravia already in possession of a fairly well developed industrial base that, despite the existence of several large industrial agglomerations in Prague and its vicinity, in north Moravia, and in north and northwest Bohemia, was noted for its considerable dispersal. The reasons for this dispersal are many, and they have not to this day been satisfactorily explained. Certainly a large part here was played by the structure of industry, in which textiles and foodstuffs were strongly represented, as well as by the scattered mining of raw materials. M. Střída[4] has pointed out that some industries had their roots in the industrialization of the landed estates, which affected, in the main, the rural townships and settlements. In addition, the existence of the old, dense network of small towns, which historically suffered little war damage, in itself stimulated the growth of small-scale industrial plants. The relatively early start to industrialization in Bohemia must also have played its part in shaping the special features of Czechoslovak settlement.

It would be a mistake, however, to imagine that there were no regional disproportions in Bohemia and Moravia. Although they were not the big macroregional imbalances known in other European countries, they did have a strong and lasting influence in shaping settlement structure. In contrast to industrially advanced areas there was southern Bohemia, which suffered long-term depopulation and lacked any significant industrial development; then there were almost the entire area of the Czech-Moravian uplands, with its poor agricultural base and sparse network of small towns, the Šumava area, and a number of smaller regions that remained outside the industrialization and urbanization processes, or that, although industrialized in the past, suffered from the economic crisis of the 1930s and its aftermath.

There was, of course, the major regional disproportion between the western and the eastern parts of the republic, which can be most briefly and pointedly characterized by the fact that in 1930 Slovakia accounted for only 10% of the country's total industrial output. With these regional disproportions, only briefly indicated here, and with the job of resettling the border regions, Czechoslovakia entered the postwar reconstruction period and started to build a socialist society.

The theories and ideas about territorial development[5] that were current in those days were responses, on the one hand, to the above-mentioned circumstances, and on the other, they expressed certain theoretical approaches. Attention was focused mainly on the problems of locating new investments, especially in industry. Decentralization tended to be the dominant theme, with further cumulation of industrial investment in areas where funds had been poured in under capitalism being opposed. Some writers recommended that workplaces should be evenly spread and that small industrial enterprises should be set up. The distribution of industry was not to be seen purely as an economic matter; "the social utility of industrial investment" should also be considered. In this spirit, under the two-year plan of 1947-48 several enterprises were established in underindustrialized regions on the periphery of the main industrial areas. But some voices were raised even in those days in criticism of even-spread theories; J. Goldmann, for instance, pointed out that it was impossible to have every district equally industrialized, and P. Hušek emphasized that the planned industrialization of economically weak regions has its limits, and that certain decisions must always be guided by economic considerations and the national interest.

In the postwar period and in the course of the First Five-Year Plan, four main lines for regional development emerged in Czechoslovakia. They can be summarized as: (1) industrialization of Slovakia; (2) settlement of the Bohemian and Moravian border regions; (3) economic development of the more backward regions; (4) further planned development and use of the existing industrial centers and agglomerations. To pursue all four aims simultaneously was extremely difficult when, in addition, the first five-year plans involved a basic restructuring of industry. The old structure, in which consumer industries predominated, was reshaped to give much greater prominence to heavy industry as the basis for economic growth. Claims on investment resources were increased even more by the transition to large-scale agriculture. These circumstances made the development of cities, settlements, and infrastructure in all the four "target" areas more difficult. Housing construction was concentrated in areas where heavy industry had been established as a foundation for future development of the economy. The housing program, which managed to provide for workers in the new industrial enterprises in the Ostrava and north Bohemian areas, was too restricted overall to permit the other aims of regional policy to be met in full. The authors of the First Five-Year Plan were aware of the large demands made by the program, and therefore in its legislative formulation it was stated that the aim was "...also gradually to equalize the economic levels of regions"; but, it went on to say, only "insofar as fulfillment of the five-year plan is not endangered." On the whole the prevailing view was, as noted in the collection on long-range changes in the distribution of Czechoslovak industry, that national requirements were preeminent and that new industries must be planned to benefit not only the regions where they were to be located but also the overall advance of the economy.

In assessing the degree to which the four main lines of regional policy have been implemented, the majority of experts agree that the greatest success was the industrialization of Slovakia. Its transformation into an industrial-agrarian country was achieved in a historically brief span, and this caused many changes in settlement structure as well. From the macroregional standpoint, in particular, the goals set in the postwar period were achieved. From the meso- and microregional standpoints the results were less satisfactory, for the general tendency to disperse investment was also evident in Slovakia. Discussing this aspect, M. Blažek writes that according to the geographical indicator for number of localities (industrial) over an area, the dispersal of industrial centers is of the same order in Slovakia as in the Czech parts of the country, if not greater.

Settlement of the border areas was on the whole completed successfully; they were integrated economically and in other respects into the rest of the country. As for the economic and, in particular, the industrial development of the more backward parts of Bohemia and Moravia, analyses of industrial employment indicate that despite the relatively large increase in the numbers employed, the proportion of the industrial work force employed in these areas remained as low in the sixties as it had been in the immediate postwar era. True, the drain from the traditional areas of depopulation had been halted, but the trend had now shifted to new regions; for instance, in the vicinity of Plzeň, the continuous zone of east Bohemia, and around Klatovy, Rakovník, and so on.

The theories prevalent at the time of the First Five-Year Plan and after it inspired measures for their implementation; among them were restrictions on the growth of population in Prague through regulation of employment there. These measures were successful; Prague was one of the few European capital cities not to grow, and, it must be emphasized, the population remained virtually stable even in the ring surrounding the city. That this success for regional policy could have its negative sides as well was seen toward the end of the sixties, when views on the growth of Prague also began to change.

One part of the drive to eliminate the most serious economic and social inequalities among regions was a move to locate new manufacturing and consumer industry capacities in the medium-sized and small towns and to expand older enterprises located there. Arguments in favor of this investment policy usually pointed to the labor reserves to be found in the places where new job opportunities were being created as well as in the surrounding areas. This was particularly true for cities in agricultural districts, where during the fifties and sixties large-scale farming and its mechanization released many people for other types of employment. The result was usually an increase in commutation to cities as well as a decrease in demand for housing, so that the existing housing stock was used, and building capacity was released for construction in areas where the development of coal mining, energy, and metallurgy were concentrated.

In preparing the first generation of the national settlement strategy, it was necessary to clarify the relation between physical and economic planning. In the period of extensive development, the relation was one-sided in the sense that the economic plan formed the basis from which specific regional solutions were derived. Physical planning merely located investments that had already been set in economic plans. In this context physical planning was described as a "projection" of economic, or even sectoral, planning. The feedback link that existed between territorial conditions, social potential, and the consequences of placing investment in a particular location and the economic plan was then not sufficiently understood. The problems caused for the functioning of cities and entire regions by this one-sided sectoral concept of planning could be tackled only when regional development was based on a good coordination of economic and physical plans.

Since development of the separate parts of a country depends on the concept for the whole and on relations among regions, in the early sixties it was found necessary to start working on the main lines for nationwide development of the settlement network.

The first generation of national settlement strategy was worked out by VÚVA (the Research Institute for Building and Architecture) at the Urban Center in Brno. The result was a proposal to categorize the prospective settlement network into a three-tier system of centers. The progressive reconstruction of the network tried "to create more balanced and also more efficient conditions for the work, living, cultural enjoyment, and recreation of the population." Underlying the idea of the three-tier system, then, was the desire to implement one of the traditional socialist principles: to eliminate the social differences dividing the various territorial units. Therefore the concept[6] stressed that the centers containing the basic components of the facilities with which their citizens and those of the hinterland zones can satisfy their rights and claims to education, health care, etc., should be spread as evenly as possible over the country. At the same time, the three-tier system should help, on the economic side, to strengthen the concentration of production and investment in the nonproduction sector.

These aims were formulated in Government Decision No. 100 on the long-range concept for developing the settlement system, issued in 1967, which states: "...the main aim should be to establish a settlement structure permitting communal and technical facilities to be concentrated in selected settlements, ensuring for all the population an optimal living environment, while at the same time having regard to all the aspects of economical construction and operation of the necessary facilities and to the requirements of production, which would also correspond to the present trend toward concentration of industrial and agricultural production in larger units."

With regard to the locating of investment, it was proposed to divide the settlements having central functions into three basic levels:

— central places of the first level, i.e., those of local significance;

— central places of the second level, i.e., of district significance;

— central places of the third level, i.e., of regional significance.

The remainder, not functioning as centers, were to be divided into:

— settlements of permanent significance;

Czechoslovakia

— settlements of temporary significance at the present time;
— the remaining settlements.

Centers of local significance were to provide people living in centers and communities within the hinterland zones with essential social amenities. Basically they would be of rural significance. They should have a minimum of 1,500 to 2,000 inhabitants, with 3,000 to 6,000 in the catchment areas. Studies estimated that in the future there would be 1,700 such centers in Czechoslovakia.

Centers of district significance are for the most part already developed towns; industry should be concentrated there, and they should provide complete facilities, serving the populations of the centers and of the hinterland zones. According to the concept, with the local centers they should form a settlement grouping providing easy access to both facilities and places of work.

Here, then, were the first definite formulations of principle for the settlement system, which was to be elaborated later. Great importance was attributed to the centers of district significance; their network was to be "as far as possible equally distributed over the settled and economically viable territory of the country." Of course, to achieve this aim meant including in this category places that were not really towns at all; in a later selection, made in 1971, communities with no more than 1,500 to 3,500 inhabitants were included. The VÚVA work assumed that the majority of these second-level centers would grow to 30,000 to 80,000 inhabitants. A concentration of this size was considered to be a sufficient base for developing all the important components of amenities and facilities. In the course of working out this strategy there was much discussion about the number of centers of district significance and about the directly related problem of their size. In the first proposals, contained in the publication Základní otázky osídlení v ČSSR [Basic Questions of Settlement in the ČSSR],[7] the figure of 251 for the whole country was envisaged. Later the theoretically projected number was reduced to 151 because doubts began to appear in the midsixties about the suitability of the rather wide investment dispersal that the first proposals would have produced. The question whether a large number of small towns or a smaller number of larger towns was needed was long a bone of contention; in the end the issue lost much of its acuity when it began to be realized that it would be better to think in terms of agglomerations, urban regions, settlement groups, etc., which made ideas about size less urgent.

The <u>centers of regional significance</u>, classed at the highest level in the settlement system, were to provide their own populations and those of broader areas with the best types of facilities and amenities, largely of a more selective and specialized character, appropriate to major centers of industrial and social life. The proposals assumed that these centers, which in addition to Prague were all the administrative capitals of new or old regions, would number nineteen. In time they should reach the 80,000 to 400,000 level of population. It is interesting that in contrast, for example, to Hungary, this three-tier classification did not place Prague in a category of its own.

The concept provided the basis for practical measures to develop the settlement system in the Czech Socialist Republic; they were outlined in the government decision on the long-range development of settlement, issued in 1971.[8] The decision stresses that "settlement development must be organized so that it is in accordance with the trends anticipated in the way of life of people in socialist society and with the economic needs and interests of that society." It also states that the strategy should ensure that "the development of settlement corresponds to the ongoing concentration of industrial and agricultural production." An important element was the selection of 170 centers of district significance in the ČSR. This was to serve as an instrument for concentrating productive and nonproductive investment, and the major part of housing construction was earmarked for the second- and third-level centers.

The research and experimental projects of this period, which looked to more distant horizons, for instance, to the year 2000 and beyond, already pointed out that there would be a greater concentration of intensive urbanization localities. The idea of some kind of urbanization belts or an urbanization skeleton for the republic also began to emerge. The VÚVA concept was effectively supplemented by "Project R," elaborated in 1964-65 by Terplan. It was actually an analysis of the main natural, social, and territorial realities that had to be taken into account in long-range settlement policies, and it was also a useful aid for selecting centers of district significance.

The proposals worked out by VÚVA, and largely applied by the government authorities, were not, however, the only theoretical works dealing, toward the end of the sixties, with the future lines for settlement. Now that we are in a better position to compare views in other socialist countries, we see the VÚVA approach at that time as recommending a strengthening of so-called medium urbanization, i.e., developing centers of district significance. There were also other proposals that either stressed belt development — along "development axes" (G. Čelechovský et al.)[9] — or, on the contrary, urged the need for decentralization and restricting as far as possible the growth of large towns (Z. Lakomý et al.).[10] Criticism directed against the large number of second-level centers and against the schematic nature of ideas on the use of facilities, the minor importance allotted to the impact of the automobile, the underestimation of concentration trends in production, and the like, led in 1970 to consideration being given in VÚVA work to alternatives that assumed, in contrast to the ideas of the sixties, greater population concentration in the regional centers. In the later discussions, which played a role in forming the views that underlay the second generation of settlement strategies, it was pointed out that the central place system failed to take sufficient account of the part played by industry in shaping the system while placing too much importance on the service function of the urban centers. Nor did it give sufficient weight to the formation of new settlement structures, such as industrial agglomerations, urban regions, and the like.

The second-generation concepts for the ČSSR were now worked out separately for the Czech and Slovak republics. Although their basic principles were the same, they differed in some respects. In line with the general urbanization strategy in the second half of the 1970s, both republics then prepared drafts for so-called regional urbanization. They were influenced by a new law, No. 50, on physical planning, dating from 1976, which placed great emphasis on prognoses for territorial units.[11] First we will describe the results of work on future settlement and urbanization in the ČSR.

The concept of settlement development and urbanization in the ČSR originated within the complex of prognoses based on the 1971 ČSR government decision; Terplan[12] worked on it from 1971 to 1975. Parallel with this prognostic study, research institutes examined relevant theoretical questions, for instance, problems of regionalization, the settlement system, the social consequences of varying degrees and forms of concentration, etc.

Typical of the period when work started on settlement forecasting was the greater weight given to concentration of population and social activities in towns than had been the case in the previous period. In part this change in the intellectual atmosphere and the shift in emphasis were caused by the criticism directed at some aspects of the central place theory as applied to the settlement strategy; in part the greater emphasis on concentration trends resulted from the changed approaches and methods used in the prognoses. There can be no doubt that the stressing of concentration was also due to the greater weight placed on the relation between settlement development and trends in industry. However, the prognoses did not reject out of hand the approved central place system; this earlier concept remained among the basic starting points.

The general approach to the prognosis,[13] which in large measure determined its results, derived most clearly from the formulation of "the chief aims of long-range development of the settlement system and urbanization in the ČSR" as stated in the final draft. Among the most important of these aims, according to the authors of the prognosis, are "...creating optimal conditions for effective location and concentration of the productive forces and nonproductive activities. At the same time, considering the need to use existing capital assets," "to create a balanced distribution of population," "optimal use of the productive and nonproductive assets, of human potential, and natural resources," "to create the conditions for...the scientific-technological revolution," "to ensure maximum opportunities for choice of place and type of residence, work, job opportunities, social contacts, access to education and culture...." Among other aims the authors included reducing loss of time on travel, providing a basis for efficient construction and organization of the infrastructure, concentration of settlement in a limited number of centers, and preserving as much open land as possible for the needs of agriculture and forestry and for recreation. Settlement development should also be in accord with demands for conservation and creation of a good human environment.

On these premises the prognosis formulated variants that in varying degree emphasized the concentration of economic and social potential in selected settlement agglomerations. At the same time, however, its authors pointed to the limits and obstacles to rapid urbanization in the ČSR. First is population growth, which varies and will continue to vary considerably between different regions but is unfavorable, especially in the Czech urban centers and their hinterlands. Most of the urban agglomerations depend overwhelmingly for their growth on migration from other parts of the country. The urbanization process will be slowed in the future as well by inertia factors in the settlement system itself, primarily by the wide dispersal of industry in hundreds of locations. Also countering concentration trends are certain changes in the way of life induced by family house building, the growth of individual and public transportation, the deterioration of the city environment, and the mounting public demand regarding environmental quality. This is expressed, for instance, in the preference shown by the people of the ČSR for living in small and medium-sized towns in communities close to cities themselves, or in more remote rural communities. In the majority of the socialist countries, the cities have greater drawing power than in the ČSR. A similar preference for small and medium-sized towns, and a rejection of cities, has been observed in the Yugoslav republic of Slovenia. The sociological interpretations of these particular preferences stress that they are typical for countries where the difference between town and country — from the standpoint of the civilization level — is no longer great.

Not least among factors limiting the concentration of population in larger agglomerations is the possible rate of housing construction, which would also have to increase with increasing migration.

Another new element in this second-generation prognosis is the view that concentration trends will no longer be toward particular urban centers but toward so-called settlement agglomerations. "As in other advanced countries, so in the ČSSR, and within it the ČSR, the process of concentration of population and activities at the highest level creates focal areas for economic and social life — development agglomerations of settlement.... They are strongly differentiated in character, in their development potentials, and in their status and functions in the wider context of the republic as a whole. They constitute the key points for development of the settlement system, for its spatial organization and its functional utilization of the whole territory."[14]

In a chapter entitled "Development-Settlement Agglomerations as the Basis for the Urbanization Process," the forecasters draw a distinction between "regional agglomerations," which form the main poles of development, and "important centers of the settlement system," i.e., lower-level agglomerations.

Regional agglomerations consist of one or more core cities, usually large or medium-sized — in Czechoslovak terms — including their immediate hinterlands. Here are concentrated the most important economic, social, cultural, administrative, and other functions. The prognosis sees the regional agglomeration system as providing the "basic key points" for development, alongside which, however, are the lower-level agglomerations, the so-called auxiliary poles of settlement. A link with the previous strategy of settlement development and with

Czechoslovakia

the system of centers of district significance is provided by the fact that both the auxiliary poles and the regional agglomerations have been selected from among the centers of district significance according to the previously established criteria. There are twelve regional agglomerations, divided into four categories. The first contains just the Prague agglomeration, with 1.6 million inhabitants. In the second are the agglomerations centered on cities or groupings of medium-sized towns; they are the agglomerations of the Brno, Ostrava, Plzeň, and north Bohemian (Ustí nad Labem, Teplice, Most, Chomutov) areas. They are of considerable economic importance, and strong growth of the tertiary sector as well as reconstruction of industry are envisaged for them. The third category includes the agglomerations of the larger medium-sized towns, namely, Hradec Králové-Pardubice, Olomouc-Přerov-Prostě-jov, and České Budějovice. These towns, too, are of considerable economic importance, but they contain a much smaller proportion of the population than the previous category. In the fourth category the prognosis places the smallest agglomerations: Karlovy Vary-Sokolov-Cheb, Liberec-Jablonec, Gottwaldov, Jihlava. The nuclei of these agglomerations are all former regional capitals, and their industrial and demographic significance as a whole is roughly equivalent to that of the third category. Table 3 gives more detailed data on all the settlement agglomerations.

The regional agglomerations are supplemented by a group of nineteen lower-level agglomerations, considerably smaller in size. They are in the nature of developmental poles in the less-developed areas. Their populations ranged in 1970 between 19,000 and 82,000. Their cores are district towns with strong industrial bases. Often their major parts consist of pairs of neighboring towns.

In 1970, 4.8 million people were living in the regional agglomerations, with 900,000 in lower-level ones. However, their importance is shown much better by the percentage of the total economically active population of the ČSR engaged in their industries: the concentration of 1,321,700 industrial employees

Urbanization in Socialist Countries

Table 3

Concentration of Population by Settlement Groups in ČSR,
1970-2000

Territorial units	Population in 1970 in thousands and %	Population in thousands in 2000 by variants and %		
		I	II	III
Regional settlement agglomerations	4,830	5,610	5,770	6,000
	49.2	51.2	52.7	55.2
Significant centers of settlement	903	1,150	1,200	1,250
	9.2	10.5	10.9	11.4
Centers of district significance apart from settlement agglomerations and significant centers	910	1,160	1,180	1,200
	9.3	10.6	10.8	10.9
Other settlements	3,170	3,030	2,800	2,460
	32.3	27.7	25.6	22.5
ČSR total	9,813	10,950	10,950	10,950
	100.0	100.0	100.0	100.0

Source: Koncepce hlavních směrů urbanizace ČSR, Terplan, 1975.

represents 63.8% of all those industrially employed.

The strategy envisages that the two categories of agglomerations will constitute the territories into which both productive and nonproductive investment will flow. And it is in the figures for the probable concentration of population that the forecast expresses these trends. Since it is impossible to make a precise estimate of all the factors that in the future will influence population distribution, the estimate for agglomeration and urbanization development is made in three variants, one assuming maximum possible concentration of population in agglomerations, the second assuming minimum, and the third making a medium assumption. Taken together, they provide an integral picture of the urbanization process in the Czech Socialist Republic to the year 2000 (see Figure 1).

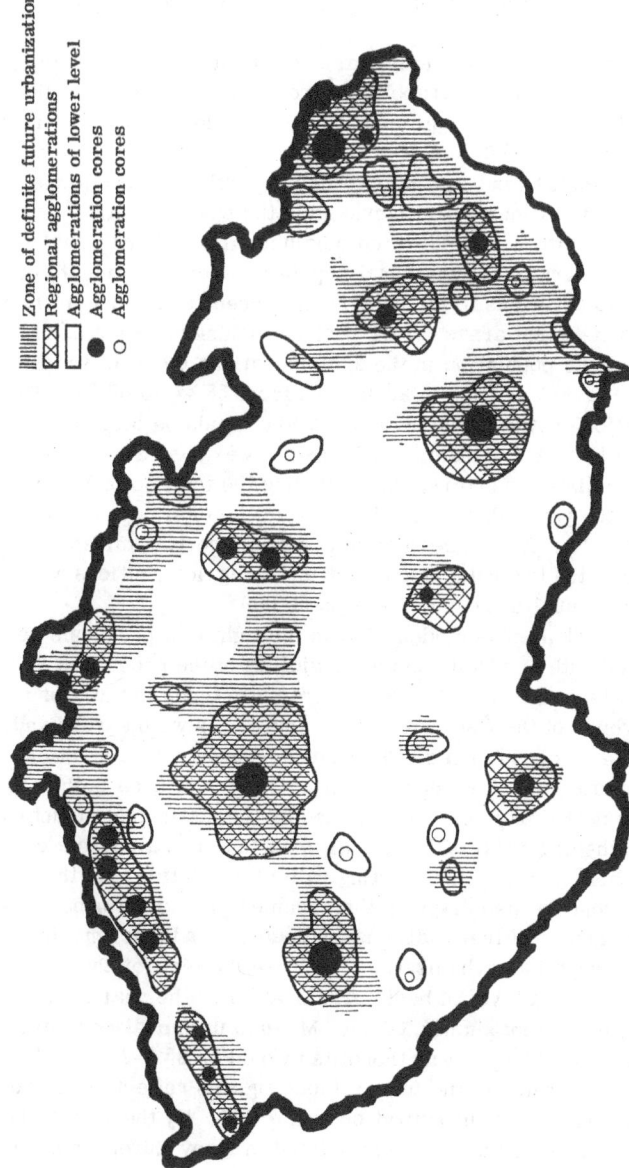

Figure 1. Main lines of urbanization in the ČSR (after Terplan).

According to the minimum variant, which the authors say is in effect a modified extrapolation from the existing trend, the population in the two agglomeration categories would rise from 5.73 million in 1970 to 6.76 million in the year 2000. Stated in relative terms, that means that the proportion of inhabitants in the agglomerations would increase in the space of thirty years from 58.4% to 61.7%. With the medium variant there would be greater migration by the rural population from the vicinity of the agglomerations to the centers of concentration, but not from the more distant areas. This variant assumes an increase in the absolute population in the agglomerations from 5.73 million in 1970 to 6.97 million in 2000, i.e., from 58.4% to 63.6%. The population growth in the agglomerations would be largest in the case of the third variant, which assumes extensive movement between districts and regions, bringing the total up to 7.29 million in the year 2000, which is 66.6% of the total population of the republic. As can be seen, there would be 1.56 million more people in the two categories, that is, the agglomerations would have an annual increase of around 50,000.

On the whole it is evident that in the light of a prognosis constructed with emphasis on the feasibility of the processes it forecasts, one cannot expect any dramatic turns in the urbanization trends of the ČSR, insofar, of course, as we measure and evaluate solely in terms of population concentration in cities or agglomerations. The moderate pace of urbanization in the ČSR for the next thirty years is dictated by the settlement structure itself, the dispersal of industrial localities, and also, to a considerable extent, by the existing rate of population growth.

The concept also displays the assumed process of urbanization by means of the traditional indicators for basic, medium, and metropolitan urbanization.[15] These suggest for the minimum variant a level of basic urbanization by the year 2000 of 62.6%; for the maximum, 66.7%. Medium urbanization would reach 48.4-52.0%, and metropolitan would be 25.7-29.8%. Although they refer to the future, these figures reflect again the special features of urbanization in the ČSR. By the year 2000 it would achieve the level that existed in many European countries in the 1960s. They also indicate that the concentration process will not have been completed even by the end of the century, and that it will continue.

Although the prognosis for the settlement system in the ČSR does not explicitly consider problems of regional differences, which in any case are not serious there, it is evident from the choice and location of the developmental settlement agglomerations that the authors intended to establish a network that would not leave even the less urbanized areas without "key points" for further growth. Nevertheless it is obvious that the southern part of the country, south of the Cheb-Gottwaldov line, has fewer settlement agglomerations than the north. The endeavor to maintain macroregional balance is also evident in the fact that the authors include among the factors limiting further urbanization the impossibility of "using" the population reserves that exist in some of the peripheral areas for the growth of cities.

Another element that should be mentioned, and which illustrates the shift in views on urbanization as compared with the ideas of the sixties, is the fact that the regional agglomerations and, to a lesser extent, the lower-level agglomerations are now regarded as the basic elements of future urbanization processes. There are thirty-one such areas, and the bulk of investments in the future should be concentrated in them. So it is no longer the district centers, with their hinterland zones, but considerably larger formations with a wider range that are envisaged as the organizers of settlement. The fact that more rapid growth of the smaller agglomerations is expected indicates that the concept is directed toward strengthening medium-sized agglomerations.

The concept above was a general outline setting the main long-range lines for urbanization as a whole. Soon, however, the need was felt to work out detailed plans in the light of further data and information from the individual regions. Therefore, immediately after the government of the Czech Socialist Republic had accepted this strategy in 1976, a start was made on drafting prognoses for the individual settlement agglomera-

Urbanization in Socialist Countries

Table 4

Structure of Settlement by Size in ČSR, 1950-2000

Community groups, population in thousands	Populations in thousands and %					
	1950	1961	1970	2000 with variants		
				I	II	III
0-5	5,245	5,100	4,680	4,090	3,835	3,650
	59.0	53.3	47.7	37.3	35.0	33.3
5-10	645	825	825	610	610	580
	7.3	8.6	8.4	5.6	5.6	5.3
10-20	590	700	805	940	995	1,030
	6.6	7.3	8.2	8.6	9.1	9.4
20-50	600	775	780	1,330	1,380	1,430
	6.7	8.1	8.0	12.1	12.6	13.1
50-100	290	485	880	1,160	1,095	995
	3.3	5.1	9.0	10.6	10.0	9.1
over 100	590	680	760	1,520	1,685	1,865
	6.6	7.1	7.7	13.9	15.4	17.0
Prague	930	1,005	1,080	1,300	1,350	1,400
	10.5	10.5	11.0	11.9	12.3	12.8
ČSR total	8,890	9,570	9,810	10,950	10,950	10,950
	100.0	100.0	100.0	100.0	100.0	100.0

Source: Koncepce hlavních směrů urbanizace ČSR, Terplan, 1975.

tions, including considering possible growth of their core towns and of the other settlements belonging to them. Such regional and more detailed elaboration of the overall urbanization concept is a specific feature of the Czechoslovak approach to urbanization strategies.

The urbanization strategy of the Slovak Socialist Republic, formulated at the beginning of the seventies, derived from the same basic principles as that for the ČSR. The rapid economic advance of the SSR, especially the industrialization rate, greater growth of population, and the morphological conditions in Slovakia made the strategy place greater emphasis on the rate of concentration trends. In past years urbanization has proceeded more rapidly in Slovakia than in the ČSR.[16]

This strategy drew on analytical and programmatic work

Czechoslovakia

that had been processed in part toward the end of the sixties. In 1967 URBION — the State Institute of Urbanism and Physical Planning — started work on its "Study of the Prospects for Development in the Slovak Regions from Territorial-Technical Standpoints." This attempted to contribute, among other things, to rational distribution of the productive forces and to creation of a long-range urbanization strategy for Slovakia. Linked to this work, which was predominantly analytical, was a programmatic study, "Main Lines for Urbanization of the Areas in Slovakia Functioning as Sectoral Centers." It assessed the current state of urbanization and the factors influencing it, while also indicating the main future trends.[17] These and other studies resulted in the drafting of "The Principles of a Concept for the Main Lines of Urbanization in Slovakia," which was adopted by the SSR government by Decision No. 95/1971. "The Project for the Urbanization of the Slovak Republic" was then formulated according to these principles. It sets out the main lines for developing the settlement system in the SSR up to 2000 and has three parts; the first deals with the macrotreatment on the scale of Slovakia; the second, with the microtreatment in the thirteen urbanization areas; and the third, with the socioeconomic and urban-legislative stimuli for further concentrating social activities and population.

In its main features the concept stems from the long-range outlook for social development in the SSR, particularly from certain circumstances,[18] for instance, that by 2000 an increment of 1.1 million people has to be accommodated in the urban settlements, and in addition, further tens of thousands must be expected to move from the countryside owing to intensified mechanization and concentration of agriculture. Population growth, migration to towns, and other requirements suggest that by the year 2000 dwellings for 1.3 million will have to be built. The growth of urban settlement calls for development of the secondary, tertiary, and quaternary sectors, with employment opportunities expanded to match. Providing for these trends also means that it will be necessary to allot and functionally and organizationally establish extensive areas for these purposes. With this prospect in view, housing and public amenities alone are expected to require 15,000 hectares.

On the basis of these forecasts, Ivan Michalec states: "It is evident that our society cannot meet these demands within the framework of the existing settlement system. This is the second time since liberation that our society has faced such grandiose tasks, requiring that we avoid all past mistakes and uneconomical measures, and that we pursue a new purposeful course while recognizing the complex Czechoslovak long-range outlook for dealing with these conditions."[19]

The principles referred to above contained a number of guidelines for future settlement development. All were aimed at the final goal, "the elimination of the contrast between town and country by creating a dynamic and open spatial and regional structure of settlement in Slovakia." To reach this goal it is necessary to observe the following main principles and guidelines:

1. To increase the urbanization dynamic in Slovakia while utilizing the favorable natural and social conditions.

2. To concentrate the natural population growth in the main areas of urbanization development, i.e., the core settlements.

3. To develop a functionally concentrated housing and industrial zone in accordance with the trends created by the main urbanization axes in Slovakia.

4. To develop urbanization along established horizontal and vertical axes shaped by the country's morphology and the concentration of its settlement.

5. To take due account of the factors which, under the specific conditions in Slovakia, lead to the development of concentric and chain forms of settlement capable of growing in the future into belt forms.

6. To achieve by the year 2000 priority development of settlement in six economic-settlement centers with over 100,000 inhabitants, namely, Bratislava, Košice, Nitra, Žilina, Prešov, Banská Bystrica, and further, in a minimum of six other spatial agglomerations with over 100,000 inhabitants and with core centers of around 50,000.

Czechoslovakia

7. To bring about an acceleration of the urbanization dynamic by means of managed urbanization at all levels of administration, so that the expanding settlement system is continuously controlled.[20]

There is clear emphasis here on stimulating and accelerating the concentration of economic and other activities and also of population, housing, and cultural activities, in selected urbanization areas. Also evident is the tendency toward developing the process in Slovakia in the form of urbanization axes intended to create in the more distant future relatively continuous urbanization zones. Obviously the geographical conditions of Slovakia play a part in shaping the axes, and the concept of developing settlement in the main valleys corresponds to the long-term trends described by Slovak geographers. The popu-

Table 5

Differences among Urbanization Variants for Slovakia

Indicators	Absolute values in		Percentage for SSR in	
	1970	2000	1970	2000
First variant				
Total population of SSR	4,542,092	5,570,000	100.0	100.0
Urbanization areas	2,800,272	3,915,000	61.6	70.3
Other areas	1,741,820	1,654,864	38.4	29.7
Second variant				
Total SSR	4,542,092	5,570,000	100.0	100.0
Spatial settlement structures of which:	2,975,761	4,285,406	66.0	77.0
urban regions	2,642,256	3,794,600	58.2	68.1
Other areas	1,566,331	1,284,594	34.0	23.0
Third variant				
Total SSR	4,542,092	5,570,000	100.0	100.0
Spatial settlement structures of which:	2,906,943	4,260,578	64.0	76.5
urban regions	3,342,788	3,466,200	51.6	62.2
Other areas	1,635,149	1,309,422	36.0	23.5

lation records for individual communities show that over the past 100 years there has been a steady decline in the number of people living at the higher altitudes and an increase in the valleys and along the main lines of communication.

If the concept is implemented, by the year 2000 about 50% of Slovakia's population should be living in centers of over 5,000 inhabitants, about 34% in communities of over 20,000 (medium urbanization), and some 20% in towns of over 100,000 (metropolitan urbanization). The urbanization process will be far from complete then, and it will continue, reaching by 2000 a point at which 65% of the population will be living in urban centers of over 5,000.

The proposals for all Slovakia, the so-called macrotreatment of the urbanization project that flowed from the "Principles" and was formulated by URBION, were worked out in three variants. All rested on common basic principles; they envisaged the development of urban zones but differed as to the degree to which the population would be concentrated in urbanization localities and in the number of urban regions or urbanization localities.[21]

Although these variants are not exactly comparable, because they state degrees of population concentration in one case for urbanization localities and in another for urban regions, it is evident that the first works on the assumption of greater concentration, and the second and third assume somewhat lesser degrees. By the year 2000 a maximum of 30% and a minimum of 23% of the population would live outside urban regions or urbanization localities — a considerable reduction compared with the present 34-38%.

Linked with the macroproject were projects for thirteen urbanization areas, i.e., the microtreatment. The basis was always the all-Slovak project, which formed the framework and was worked out in greater detail, as were the regional urban programs in the ČSR.

·2

THE SOVIET UNION

The Soviet Union is a country which, from the standpoint of urbanization and the concentration of population in cities, has experienced in a relatively short span exceptionally large changes and which deserves special attention.

Within its fifty-year existence the Soviet Union had achieved by the 1960s world priority in the number of cities with over 100,000 inhabitants, and during the first five-year plans it reached the highest urban concentration rate ever recorded before World War II in any major country. Although the high growth rate of 1926-39 declined in later years, the figure has remained remarkably high throughout the postwar period and to this day. As a result, by the midseventies 60% of the total population of the USSR was living in urban communities. An illustration of this rapid urbanization can be provided by data on the growth of some industrial cities and administrative towns with large industrial concerns.[1] Sverdlovsk, Cheliabinsk, and Perm were small places before the October Revolution, although they had been founded in the eighteenth century. By 1926 Sverdlovsk already had a population of 140,000, and by 1974 it was an industrial metropolis with 1,122,000 inhabitants. Cheliabinsk had a population of 59,000 in 1926; by 1974 the figure was 947,000. Perm numbered 120,000 in the mid-1920s; today it is 920,000. Similarly Minsk, although the capital of the Byelorussian Soviet Republic, had a mere 132,000 inhabitants in 1926, whereas today its population is about one million.

Urbanization in Socialist Countries

Table 6

Urban and Rural Population of the USSR,
1897-1975

Year	Total population in thousands	Population in thousands		Percentage population	
		urban	rural	urban	rural
1897	124,649	18,436	106,213	15	85
1917	163,000	29,100	133,900	18	82
1922	136,100	22,000	114,100	16	84
1929	153,411	28,733	124,678	19	81
1937	163,772	46,636	117,136	28	72
1939	190,678	60,409	130,269	32	68
1950	178,547	69,414	109,133	39	61
1955	194,415	86,261	108,154	44	56
1960	212,372	103,618	108,754	49	51
1965	229,628	120,730	108,898	53	47
1970	241,720	135,991	105,729	56	44
1975	253,261	153,110	100,151	60	40

Source: Naselenie SSSR 1973, Statisticheskii sbornik, Moscow, 1975, p. 7.

Similar growth was recorded by Novosibirsk which, because its development started only after the opening of the Trans-Siberian Railway in 1893, remained a small town, topping the 100,000 mark only in the twenties; today, however, with its population of 1,243,000, it is among the ten largest cities in the USSR. The growth of Alma-Ata, Gorky, Donetsk, Zaporozh'e, Dnepropetrovsk, Kazan, Kuibyshev, and many other towns has been equally tempestuous, demonstrating the speed with which industrialization and the associated concentration of people in urban centers have proceeded in the Soviet Union. The majority of Soviet towns fall into the categories of mixed administrative centers (centers of various political and economic territorial units) and industrial towns.

Let us now look, at least briefly, at the history and stages of urbanization in the Soviet Union and in prerevolutionary Russia. Up to the middle of the nineteenth century, the growth of urban population in tsarist Russia was small. Between 1811 and 1867 the annual increment was 1.5%. The beginning of industrializa-

The Soviet Union

Table 7

Selected Examples of Rapid Growth of Soviet Cities, 1926-74

City	Population in thousands for				
	1926	1939	1959	1970	1974
Alma-Ata	44	222	456	730	813
Donetsk	174	474	708	879	934
Zaporozh'e	54	289	449	658	729
Kazan	179	406	667	869	931
Kuibyshev	176	390	806	1,045	1,140
Leningrad	1,737	3,401	3,340	3,987	4,243
Minsk	132	237	509	917	1,095
Moscow	2,080	4,542	6,044	7,077	7,528
Novosibirsk	120	404	885	1,161	1,243
Perm	121	306	629	850	920
Sverdlovsk	140	423	779	1,025	1,122
Cheliabinsk	59	273	689	875	947

Source: Naselenie SSSR 1973, Statisticheskii sbornik, Moscow, 1975, pp. 26-35.

Table 8

Growth of Cities in the USSR*

Cities	1897	1926	1939	1959	1970
With more than 100,000 population	14	31	82	148	221
Population in millions	4.4	9.5	27.0	48.6	75.6
Average size of cities in thousands	314	306	329	328	342

*Census figures and in borders as of census year.

tion and the building of railroads in the second half of the century increased the tempo between 1867 and 1913 to 2.3%. World War I and the Civil War reduced the urban population from 18 to 15% of the total population; from 1921 onward, however, urban concentration continued. Between 1920 and 1926 the average annual increase of urban inhabitants reached 3.7%, higher than prewar. Urbanization reached dramatic proportions in

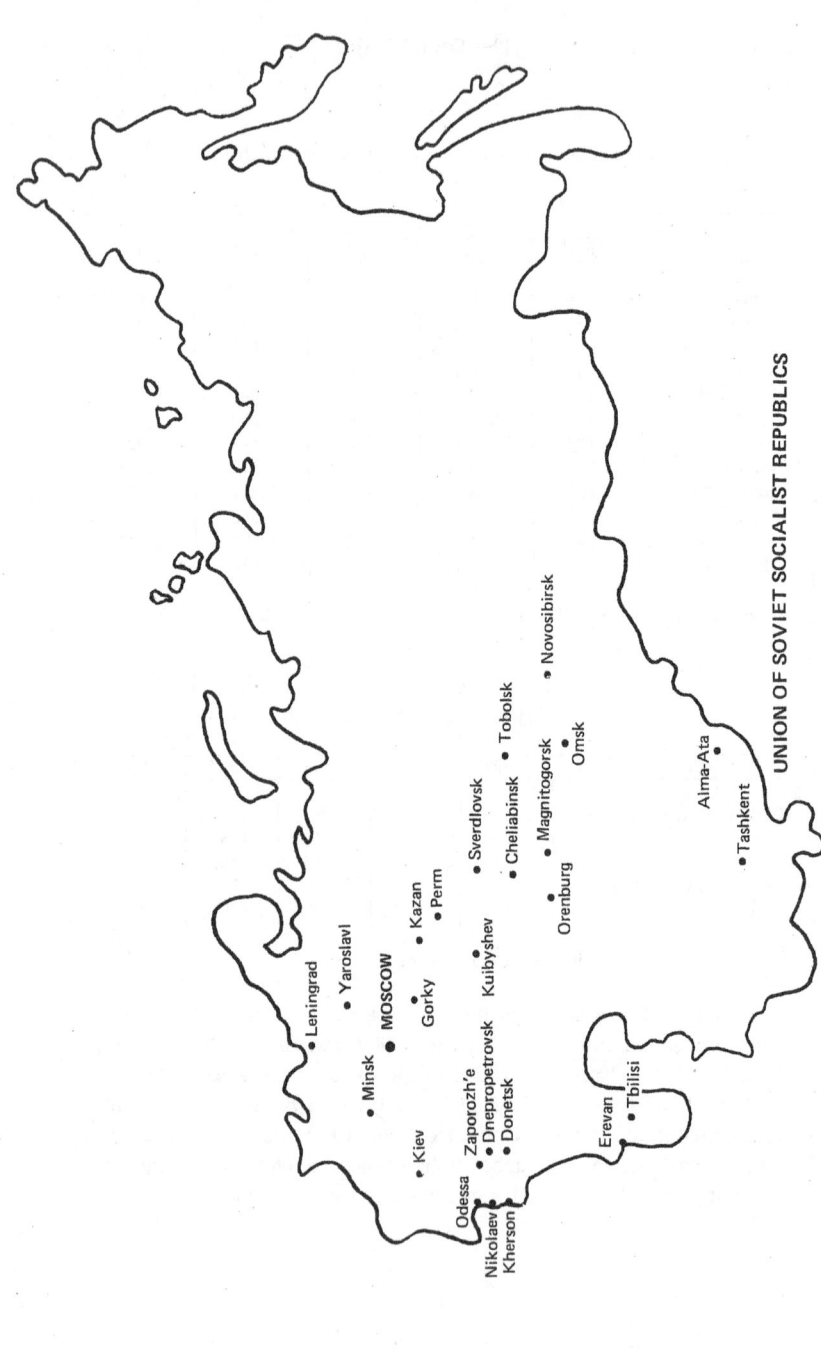

1926-39, the annual growth averaging 6.5%. But the dimensions of the process are better revealed by two figures: in 1926, 26 million people lived in urban communities; by 1939 the number was 60 million. World War II caused not only the loss of millions of lives but also the destruction of most cities in the western section of the USSR and a drop in urban population. The effect of the war was seen in an annual increment of only 1.3% in the 1939-50 period. After 1950 the urbanization rate again increased, the annual growth rising to 4.1% between 1950 and 1959. A decline started after 1959, due mainly to the already fairly high level of urbanization — roughly half of the total population lived in cities — so that urban growth had to come more and more from "internal sources," i.e., natural population increases.

A feature of the post-1959 period is, however, that 88% of the total population increase came from cities with over 100,000 inhabitants. That is to say, it was a period of urban growth by cities and their agglomerations.

In the 115 years between 1811 and 1926, the proportion of urban population rose by only 11% (from 7% in 1811, to 18% in 1926), whereas the 1926-75 period saw an increase of 42%. In 1926 there were 26 million people living in urban centers; by 1975 the figure was 153 million. Within fifty years the Soviet Union changed from a rural to a predominantly urban society, with the transformation in some areas reaching such a level that one can describe them as having a typical urban sociospatial structure. This applied in particular to big industrial agglomerations and urbanized regions, which in recent years have attracted the intense interest of Soviet planners, geographers, and economists.

The new forms of urbanization represented by the agglomerations and urbanized regions are developing quite rapidly, as can be seen from Table 9, which compares the situation in 1959 and 1970. The agglomeration processes, which in any case were already in motion, are seen by contemporary Soviet writers as manifestations of higher forms of urbanization corresponding to socialism in its developed form; in the future

Urbanization in Socialist Countries

Table 9

Development of Ten Largest Urban Agglomerations in the USSR, 1959-70

Center of agglomeration	Population in		Center as % of agglomeration	Population in		Center as % of agglomeration
	agglomeration center in thousands	other settlements in thousands		agglomeration center in thousands	other settlements in thousands	
Moscow	6,009	2,366	72	6,942	3,575	66
Leningrad	2,985	597	84	3,515	834	81
Donetsk-Makeevka	1,387	997	58	1,606	1,129	59
Kiev	1,110	186	86	1,632	279	86
Gorky	941	508	65	1,170	662	64
Tashkent	927	168	85	1,385	320	82
Kuibyshev-Togliatti	806	342	71	1,296	376	77
Sverdlovsk	779	416	65	1,025	477	68
Kharkov	953	235	80	1,223	239	85
Dnepropetrovsk	661	410	62	862	492	64

Source: Iu. L. Pivovarov, Sovremennaia urbanizatsiia, Moscow, 1976, pp. 136-37.

they will become continuous areas of intensive urbanization. This view has been summarized by Iu. L. Pivovarov: "Typical of developed socialism is the transition to an intensive mode of urban development, the spread of interconnected centers, increase in the influence of towns on their rural surroundings, removing the boundaries between towns and their surroundings — all this creates the objective conditions for the maturing in the USSR of new, more complex forms of settlement. They are primarily cities and great urban agglomerations based on them which, in our view, are becoming the focal points for spatial development and constitute the cores of urban regions and zones."[2]

Examples of these urbanized regions are the central Moscow region and the Donets, Volga, and Urals regions. The biggest is the Central Economic Region, which was formed on the basis of the Moscow agglomeration. Of its 28 million population, 20 million live in towns. The other urbanized areas are smaller, but even the smallest have populations of over 10 million.

Most contemporary Soviet writers interpret agglomerations as belonging to so-called group forms and systems of settlement that are replacing the traditional network of separate, more or less independent cities. It can be said that intensive study of how the existing agglomerations have taken shape has led to the concept of group forms, which has become the main programmatic theory behind planned urbanization.

In comparing the urbanization process in the Soviet Union with that in other socialist countries, the use of data for the whole USSR leads to simplification. The territory, which is larger, for instance, than the South American continent, includes widely differing areas. The differences lie in natural conditions, settlement structures inherited from the past, varying degrees of industrialization, communications, as well as ethnic and demographic structures and processes. Consequently, we need to divide the country into areas when examining the settlement structure.

Divisions can be made in many ways, e.g., along political lines, from the viewpoint of economics and planning, even by climate. Best suited to our purpose is the zoning worked out by B. S. Khorev, which is based on differences in the character of settlement. Essentially it also involves a typology of urbanization levels and forms covering large areas of the Soviet Union.

Khorev[3] distinguishes three settlement zones (regions), which he then subdivides. The first, which he terms "the main settlement belt," is divided into the western zone, embracing the European part of the USSR with the Urals but excluding the northern and southern territories; the middle zone, comprising the territory along the Trans-Siberian Railroad and northern Kazakhstan and the Altai; and the eastern zone, covering the southern regions and the autonomous republics of eastern Siberia and the Far East.

Second is the southern zone, divided into the Caucasus region and Central Asia. The third, northern zone has two parts — the European and the Asian north. From the standpoints of density, changes in density, level and rate of urbanization, and last, the nature of the urban networks, Khorev defines the following types of settlement zones:

1. Western zone. Features high population density (10-50 per km^2), low population growth, relatively high urbanization level, below average rate of the urbanization process, developed urban network.

2. Caucasus region. High population density (50-70 per km^2), high population increase, urbanization below the national level, high rate of the urbanization process, and a fairly developed urban network.

3. Central Asian region. Marked by a medium population density (10-20 per km^2), rapid population growth, urbanization at a low level but advancing very fast, urban network fairly well developed but with small towns predominating.

4. Middle zone. Low population density (3-7 per km^2), above average increase, urbanization at fairly high level and proceeding at a medium rate, urban network unbalanced with strong concentration in the large towns.

5. Eastern zone. Low population density and low increase, level of urbanization high but proceeding at below average rate, urban network insufficiently developed.

6. The European north. Low population density (1-3 per km^2), average increase, but proceeding at below average rate and with the urban network insufficiently developed.

7. The Asian north. Very low population density (0.1-1 per km^2), high increase but, owing to the enormous area, density has not substantially increased, level and rate of urbanization fairly high, urban network undeveloped.

The above typology, based on analysis of the current situation and trends, though very instructive, is not sufficient for an understanding of the specific features in the structure of settlement in the various regions of the Soviet Union nor for comparison with settlement systems in other socialist countries.

The Soviet Union

The classification and typology need to be supplemented, albeit very briefly, by some notes on the historical development of the urban network in the USSR.

Settlement and urbanization processes have developed in the Soviet Union on the one hand from the old urban and rural settlement pattern, on the other by the building of new towns. To a certain extent this distinction is relative because many Russian cities that we regard as "historical" are of fairly recent origin. This applies especially to cities in the Urals, Siberia, and the Far East. Nevertheless the distinction is justified because in no other period of Russian history have so many towns originated as in the period following the October Revolution of 1917. Between 1917 and 1969 a total of 1,034 new towns appeared, and in about three quarters of them, i.e., in 751 cases, the chief factor in their origin was industry.[4]

The oldest parts of urban settlement in Central Asia, the Caucasus, and on the shores of the Black Sea date back to medieval times; the origin of the Slavic towns begins with the ninth century. After the defeat of the Tatars, the political power of the Russian state spread gradually to the east, southeast, and north of Moscow, and with it went the growth of Slavic towns and rural settlements. The main urban area, although with very small towns, remained for centuries, however, to the west of the Volga River. Then, in the second half of the nineteenth century, came the rapid growth of the Black Sea ports (Odessa, Kherson, Nikolaev), in response to the growth of international trade, and the growth of the old towns of Tashkent and Tbilisi to the southeast. During World War I these twenty largest towns were joined by two more, Orenburg and Omsk, lying to the east of the Volga. The distribution of these twenty cities is an indirect indication that the focus of urban settlement in the prerevolutionary days lay in the central Russian area, the Ukraine, northwest Russia, and the Baltic and Caucasian regions. The urban network in this area consisted both of administrative centers, mainly the capitals of guberniia, and of industrial towns, ports, and industrial centers of transportation. They were fairly evenly, but not very densely spaced.

Urbanization in Socialist Countries

On the whole one can say that in comparison with the Czech lands, Poland, Croatia, and some other Central European areas belonging to the socialist countries, this historic territory was less densely settled; but its individual communities, including the rural ones, were larger. In the other parts of tsarist Russia, the network of urban and rural settlement was much less developed; it was concentrated along rivers and railroads, and especially in Central Asia it matched the economic conditions of arid regions, lacking industry and modern agriculture.

After the revolution rapid and intensive industrialization, with its accompanying urbanization, was centered, in different periods, in several areas. In the European part of the USSR these were, before World War II, the Donets Basin, the central industrial region including Moscow and Gorky, Leningrad, and the industrial towns on the Volga. In addition there were the Urals industrial area, the Kuznets Basin, and towns in Siberia and Central Asia. In the period from 1939 to 1959, which included the war years, the focus of urbanization shifted largely east of the Volga, to western Siberia and to Central Asia,[5] whereas after 1959 the rapidly growing cities were more evenly distributed between the western borders of the USSR and Lake Baikal, with Kazakhstan and some other Central Asian areas still among the regions noted for rapid growth.[6] According to Khorev's zoning, the most rapid growth in urban population during the period between the 1959 and 1970 censuses was in the entire southern zone, while the increase in Central Asia (56%) was higher than in the Caucasus (47%). The rapid urban growth in the southern zone is in line with the general macro-regional trend of population distribution in the USSR, which shows an increase in the relative share of population there and a noticeable decline in the proportion living in the Russian Federation and in the greater part of the Ukraine.

A specific feature of settlement in the USSR is, then, the marked variety of structure manifested in the varying degrees of development and maturity in the rural and urban networks, so that one cannot really speak of a single, uniform system. The settlement and urban network is on average less dense than

The Soviet Union

Table 10

Percentage of Population by Settlement Categories in the USSR, 1926-70

Settlement category	Year			
	1926	1939	1959	1970
Rural communities	82.1	68.3	52.1	43.7
Towns under 50,000 inhabitants	8.6	13.1	19.3	19.5
50,000-100,000 inhabitants	2.8	3.7	5.3	5.4
100,000-500,000 inhabitants	3.7	8.2	11.7	15.9
Over 500,000 inhabitants	2.8	6.7	11.6	15.5

Table 11

Growth of Towns and Urban Population in the USSR, 1926-1970

Urban communities in thousands of population	Number of towns, 1970 as ratio of 1926	Population, 1970 as ratio of 1926
10	136.9	147.1
10-20	301.1	303.8
Total to 20	197.8	241.9
20-50	437.0	455.3
50-100	313.3	317.1
Total 20-100	397.3	383.5
100 +	712.9	795.8
of which over 500	1,100.0	909.8
Total USSR	272.9	535.9

Source: M. L. Strogina, Sotsial'no-ekonomicheskie problemy razvitiia bol'shikh gorodov v SSSR, Moscow, 1970, p. 14.

in the other socialist countries, and this is true for the greater part of the European regions. The Soviet Union is a country of large cities, as is evident, for example, from the proportion of the urban population in cities of over 100,000.[7] Urbanization occurs by enlarging old town centers, but also to a large extent by the building of new towns. This mode of urbanization plays a much stronger part than in any other European socialist country. Another characteristic is the existence alongside the cities of many so-called urban settlements inhabited mainly by industrial workers. Industry and administrative functions are the prime factor in Soviet urban development.

The strong concentration of the industrial population in urban centers and the relatively minor extent of commuting are in large measure the result of low population density and the great distances separating settlements in the peripheral areas from their urban centers. This does not apply, however, for the industrial agglomerations, which are noted for high density of settlement and population.

Although the settlement structure is notable for the large number of cities, medium-sized and small towns are most numerous; and according to some Soviet writers, one can see in some of them a crisis in their function. All the above features of the settlement structures are then reflected in the settlement strategies.

Settlement Strategies in the USSR

Changes in settlement had been influenced by planning in the Soviet Union even before World War II. Settlement strategies, although the term was not yet in use, grew from the theory of socialist location of the productive forces and within the framework of Soviet town-planning theory. In contrast to the earlier attempts by socialist writers to formulate ideas, the concepts were now directly connected with the reality of constructing a new society. We cannot deal here in detail with all the views held by Soviet writers during the interwar years regarding urbanization, but it is essential that we at least briefly mention the main theoretical standpoints that gradually emerged in both fields. As we shall see, they dealt with issues that even today are far from closed or definitively resolved.

Since the theory of socialist location policy had considerable influence on the building of towns and on urbanization throughout the postrevolutionary period, it will be useful to pay attention to the main phases of its development. Following a brief period when some Soviet writers tried to apply Weber's principles to the location of industry and regional economies, the close of the twenties saw the emergence of a specifically Soviet theory. Its starting points were the ideas in the works of Marx, Engels, and Lenin about urban development, the relation between town and country, and industrial development.

Another important source was the experience of building socialism in the USSR as expressed in decisions of Communist Party congresses and the party's Central Committee. Along with the principles already formulated in the works of Marx and Engels, which, with the aim of eliminating differences between town and country, recommended a more even distribution of industry and population and saw various combinations of agriculture and industry as the way to achieve this, the main inspiration came from Lenin's "Draft Plan for Scientific and Technical Work," written in April 1918. In setting out the main principles for reorganizing industry, Lenin wrote that the plan should include "the rational location of industry in Russia from the standpoint of proximity to raw materials and the lowest consumption of labour-power in the transition from the processing of semi-manufactured goods, up to and including the output of the finished product; the rational merging and concentration of industry in a few big enterprises from the standpoint of the most up-to-date large-scale industry, especially trusts...."[8] Lenin dealt with problems directly or indirectly concerning the distribution of industry on other occasions as well, especially when the GOELRO plan was being drafted, and the first work was being done on the state plan.

At first some Soviet economists saw a contradiction between the idea of spreading industry evenly and locating it close to the sources of raw materials; they supported Lenin's idea of bringing industry to the raw materials, which would benefit the economy more than if the "even" distribution theory were applied. Others, who tried to take a synthetic view, believed that promoting the dispersal of industry would correspond to the aims of socialist society; but it would have to be understood that not all types of industry could be distributed evenly over the country, and that the location of many other sectors was strongly dependent on sources of raw materials — it was therefore rational to site factories close to these resources.

Soviet regional strategy and the territorial development of industry proceeded, in accordance with the actual historical circumstances, along both these lines. Party congress decisions in the twenties recommended a more even distribution of industry (Fourteenth Congress in 1925) and industrial development in the less advanced regions (Fifteenth Congress, 1927), but taking into account countrywide economic needs. This approach was also in tune with the party's nationality policy, a factor certainly influencing Soviet policy on settlement. At the Tenth Congress in 1921 it had been emphasized that the relationship between the European and Asian parts of tsarist Russia could be defined as that of a metropolis to its colonies. The prime aim of the proletarian revolution was therefore to remove all the remnants of national inequality from all sectors of social life, especially in the economy, by means of a planned redistribution of industry to the more remote areas and by moving factories closer to raw materials. In practice these principles led to the rapid growth of cities throughout the area east of the Volga. In the thirties, in addition to the emphasis on developing an industrial base in the east, especially east of the Urals, a further aspect appeared in the decisions of the Seventeenth Party Congress. This was the recommendation that light industry and foodstuffs be situated close to their widespread sources of raw materials, which would help to industrialize areas unsuited to heavy industry. By that time the strategy now known in some socialist countries as polycentric concentration was in effect being applied. Naturally there were plenty of problems in interpreting and applying the idea of "even distribution of industry and population." In some interpretations, especially among town planners, it was seen to mean dispersal, a radical decentralization of settlement, "the abolishing of cities" in the spirit of the theory advanced by B. Taut, who worked as an architect in the USSR during the twenties and thirties.

This idea acquired a different meaning in light of the industrialization carried out under the first five-year plans and the associated theory of location of productive forces. As E. Goldzamt wrote: "...in the first phase of industrialization, that is, in the phase of building heavy industry, it was not possible to disperse new industries. The construction of big plants led to the growth of urban centers. And for the avant-garde the dimensions of this process were incredible. After all, we cannot transfer 100 million Russian muzhiks into towns, wrote Ginzburg in a letter to Le Corbusier, polemicizing against the tendency to restrain deurbanization. Thirty-five years later the towns and cities of the USSR had 100 million more inhabitants than before the start of industrialization."[9]

The dominant and long-term influence on town building, on shaping the relationship between town and country, and on Soviet urbanization came from a resolution passed at the plenary meeting of the Communist Party's Central Committee in 1931, which considered this whole range of questions. Before mentioning some of its recommendations, we will sketch the development of Soviet thinking on town planning in the twenties. Otherwise some parts of the resolutions would be incomprehensible.

While in the context of regional economics concerned with the location of economic activities, the main source of discussion and conflicting views had been the degree to which productive forces should be concentrated on the macroscale of the entire country or in the regions, in the context of town building and town planning the question was the degree to which settlement should be decentralized. Soviet town planners, basing themselves on Marx and Engels's ideas about the even distribution of industry and population, worked out after 1918 a number of plans for Soviet cities — Moscow, Leningrad, Baku, Yaroslavl, Erevan — and then for the cities involved in implementing the GOELRO plan. The central idea, especially in the city plans, was active deglomeration. Thus the "new Moscow" concept of 1919 rested on a decentralizing strategy and proposed the formation of several satellites. The 1924 plan was also decentralizing in character.

Decentralization was also the theme of later theories that originated in the second half of the twenties and were the subject of learned discussions among their proponents. The first, which presented the model town of the future as based on "urban communes" and "agrotowns," suggested forming towns of 50,000 to 60,000 inhabitants who would live in 50 to 100 large collective blocks of flats. The spread of such places in the countryside was intended to eliminate the differences between town and country. In farming areas there were to be large production units whose workers would be gathered into "agrotowns," also of about 50,000 inhabitants. So, on the one hand, decentralized industry and urban settlement, on the other, concentration of agriculture and rural settlement — with the aim of removing the disparity between urban and rural living. This meant the disappearance on the one hand of the cities, on the other, of the small rural communities, accompanied by the emergence of a settlement network composed in the main of medium-sized towns.

Opposing this urban theory was the idea of scattered socialist settlement, settlement belts, as documented in the proposals for the regional development of Magnitogorsk, Moscow, and other cities. It was more radical than the first concept because it aimed at maximum dispersal and at distributing the population and their activities as evenly as possible. The basis was to be decentralization of three elements in production — energy, raw materials utilization, and processing. The supporting facilities were to be brought as close to the users as possible.

The 1931 Central Committee resolution rejected the decentralization theory and recommended that the town-country gap should not be eliminated by abolishing cities but by transforming them, while simultaneously carrying out a socialist transformation of the countryside, bringing it under the influence of progressive urban culture. The resolution also said

The Soviet Union

that industrial construction must in the future be directed to establishing new industrial centers in agricultural areas, which would ultimately bring the day when the gap between town and country would be closed. At the same time, it was decided to stop building new plants in Moscow and Leningrad. At the Eighteenth Congress in 1939, the ban on new industrial construction was extended to five more industrial cities in the western part of the Soviet Union. This was a further stimulus to locate new industries in the developing areas to the east and to establish so-called medium-sized cities. The majority of them formed the core for new and extensive industrial complexes that represented a new type of settlement. The experience of planning them (when completed, their populations numbered several millions each) was a step toward the settlement strategies that were worked out after World War II.

Today the basic theoretical pattern for developing both the individual elements and the system of settlement in the USSR is the "group settlement system." But before this concept matured in an integrated form, a number of important studies and proposals appeared. We must make at least brief mention of them.

When the period of postwar construction was over, the Soviet republics began to analyze their settlement networks and to draft proposals for development. Similarly, analyses were made for the big industrial regions, e.g., the central economic region of the Russian Federation and the Kharkov-Dnieper Region. They can now be seen as stages in forming the theory of the group settlement system, which appeared in its final version at the beginning of the seventies.

Other proposals and studies of settlement systems in the union republics yielded a number of themes tending toward conclusions at variance with the principles of the group system. We can mention, for example, a series of studies that revived the belt settlement model. E. Goldzamt points out that some of the proposals recall the "scattered settlement" scheme of 1930.[10]

Some problems and themes have constantly recurred ever

since the war. They form a core of questions that every approach to settlement in the USSR has to take into account. In particular, they are questions concerning the growth of large cities and the reconstruction of rural settlement. In addition new development forms are sought for the highly urbanized areas and settlement groups. Interest has also constantly centered on ideas for creating special types of settlement in areas newly opened up, for instance, the virgin lands and western Siberia.

The policy of concentrating large industrial plants in the cities, which has been pursued since the thirties and has continued in the postwar era, is regarded as correct; but some Soviet economists have voiced fears that concentration may now be excessive. For instance, R. Zverev[11] has pointed out that roughly one third of all industrial plant is located in cities with more than 500,000 inhabitants. And new investment continues to flow primarily to these centers.

Concern about excessive city growth was also evident in the search for the optimum size of cities, to which a number of authors devoted great attention, especially in the fifties and sixties. Today it is seen as impossible to fix an optimum size independently of the city's functional and regional role. The estimates of optimum size, made for the most part in the late fifties and early sixties, varied. According to V. G. Davidovich the optimum ranges between 100,000 and 500,000, A. E. Probst puts it at 50,000 to 100,000; studies of the physical planning of districts and nodal areas put the figure between 30,000-50,000 and 300,000; I. Bocharov et al. gave a range of 180,000 to 250,000; A. Skvortsov postulated 100,000 to 200,000; and K. F. Kniazev, 50,000 to 100,000.[12] The majority of these optima were below the limits of many Soviet towns and cities; this stressed, in effect, the advantages of so-called small cities. The arguments advanced by the authors of the optimization theories contributed to supporting the growth of medium-sized urban centers, primarily with 50,000 to 250,000 inhabitants, and this was reflected, for instance, in the planning of settlement in the Ukraine. Activization of small and medium-sized towns was among the most frequent injunctions at that time.

Soviet studies of the fifties and sixties repeatedly stressed that the main problem for the settlement network was the "fragmentation" of rural settlement. In the 1950s there were some 700,000 rural communities, nine tenths of them inhabited by fewer than 100 people. At the time about 40% of the entire rural population lived in such places. However, since then the countryside has experienced major changes, manifested in the realm of settlement by rapid concentration of population in larger communities, while hundreds of thousands of small settlements have disappeared. V. S. Riazanov, one of the foremost experts on rural settlement, states that according to the 1959 and 1970 censuses, the number of rural settlements dropped from 704,800 in 1959 to 469,300 in 1970. "In this process the number of small rural communities of up to 500 inhabitants declined by over one third, and today 44% of the total rural population is living in settlements of over 1,000 inhabitants."[13] The trend toward concentration of rural settlement was undoubtedly strongly supported by the merging of kolkhozes that began in 1951-52; moreover, it has been put into effect in all regional planning projects and in constructing numerous experimental rural settlements. All the proposals for rural settlement development in the union republics also tend toward strong concentration. For instance, in the Ukrainian Soviet Republic, which in the sixties had some 100,000 rural communities, only 10,000 kolkhozes and 1,000 state farm settlements are envisaged for the future. In Byelorussia proposals for rural settlement involve reducing the 34,000 small communities to 4,000, with these to be divided into three categories. A similar development is scheduled for the network of rural communities in Lithuania and in other parts of the Soviet Union.

An important element in the rural settlement strategy is the establishment of industrial complexes and combines. They depend for their formation on industrial methods being introduced into agriculture, on developing associated industries, and possibly on small factories being opened in rural communities. This trend toward a new type of settlement, the so-called agroindustrial type, is vital for the urbanization of rural areas in the USSR that have low and medium population densities. Under present conditions daily commuting to industrial cities is in some parts of the country practically impossible, yet the greatest labor reserves are to be found in the rural areas. Moreover, with farming itself industrialized, with farms merged and units of 10,000 to 20,000 hectares being formed, plus the concentration of farmers and people employed in services, it becomes possible for settlements of up to 10,000 people to be established. Combining farming and service functions with small-scale industry would promote the urbanization trend in the Soviet countryside. Riazanov considers that "...perfecting the organizational forms in agriculture and intensifying the links in the fields of production, science, and public amenities among kolkhozes and communities will promote a growing tendency for local settlement systems to develop around the small towns that are district centers and around places that are becoming the centers of agricultural enterprises...."[14] A whole series of proposals for these rural settlement structures was advanced during the sixties, while there was also a revival of the agrotown idea dating from the late twenties.

A special problem of importance for the Soviet Union is connected with establishing settlements in newly opened territories. These areas differ from each other considerably, and the functions of the new settlements are also quite varied. Farming settlements on the virgin lands are very different in character from the "expedition" towns in the far north, which usually serve the people working in mining projects. In the majority of these territories the tendency prevails to concentrate population and services. In some areas there is no point in building anything but towns. The well-known Soviet town planner B. Svetlichnyi had stated quite clearly in the sixties how some of the untouched parts of the country should be settled: "It is not, after all, a matter of abstract theory but a purely practical job. What settlement system should be used,

for instance, for the West Siberian lowlands? A shapeless
structure of small units beside each group of oil wells in the
midst of endless marshy forests, or several large towns linked
with the extraction sites by a network of rapid transportation,
e.g., on the principle of cableways? Obviously the second is
much more suitable...."[15] In fact, this second method is more
commonly used in the theory and practice of dealing with the
undeveloped areas, and a number of new towns have been established in this way. This also explains why the northern
European and Asian parts of the Soviet Union and the Far East
are among the most urbanized areas of the country. In these
sections 70% of the population lives in towns, a figure identical
with the level in the most highly urbanized countries.

Table 12

Reasons for Origin of New Towns in the USSR,
1917-69

Main factor	Number of new towns	
	absolute	in %
Industry	751	74
Transportation	64	6
Administration	201	18
Spas	18	2
Total	1,034	100

Source: I. P. Muravev and S. V. Uspenskii, Metodologicheskie problemy planirovaniia gorodskogo rasseleniia pri sotsializme, Leningrad, 1974, p. 50.

Although the proposals and directives for activizing medium-sized and small towns, for restricting city growth and establishing a more concentrated settlement network in the countryside, and for planning the settlement structure in the new territories have contributed to solving many practical planning matters, they have not been sufficient when it comes to the complicated job of planning and managing entire networks of settlement.[16] Along with many progressive processes connected with rapid urbanization in the USSR, there remain some aspects in the development of the settlement network that make the planned management of urbanization difficult. Belousov writes: "The historically developed system of settlement is still marked by unequal distribution of population over the country, by the existence of two settlement systems that are not yet sufficiently linked with each other — the urban and the rural — by unequal development of towns and settlements of different sizes, by fragmentation or scattering of rural settlements.... The development of urban agglomerations, based on the largest and large towns, is still under only weak townplanning control and has been accompanied by a number of undesirable economic effects. An autonomous approach to urban development still prevails, without regard for the relationships actually existing in economic life and in employment and public amenities among whole groups of settlements. Insufficient attention is paid to the progressive process of opening branches of enterprises, workshops, and auxiliary industries attached to large firms in the small and medium-sized towns that lie in the hinterland zones of large cities."[17]

We have quoted at length an author who heads an institute responsible for directing research concerned with working out principles for settlement development because he summarizes the difficulties that a new settlement theory has to overcome. They stem in large measure from the lack of a link between the activities of the various sectors and physical planning or a link between the regional and physical plans for different parts of the country. Such complicated structures and processes can only be controlled if systems procedures are fully employed both in analyzing settlement and in modelling and then managing it. The theory of the group settlement system, which tries to do this, is at present the most fully elaborated concept in the Soviet Union. Other theories are undoubtedly of less theoretical and practical importance.

The most recent version of the group settlement system has been formulated in "The Basic Theses for a General Scheme of Settlement on the Territory of the USSR," which is the product of several years' research and experience in the planning of cities and, especially, of large economic areas; it is also undoubtedly the product of the broad exchange of views on urbanization problems that has been taking place in the Soviet Union in recent years. In addition to the official document, which is a synthesis of the views arrived at by a group drawn from the Central Research and Design Institute of Town Planning, similar institutes in Kiev and Leningrad, the Sociological and Geographical Institutes of the USSR Academy of Sciences, and other organizations, a number of other studies have been published that provide, together with commentaries on the main themes, full and reliable information which allows us to summarize the main points of the theory.[18]

The studies containing settlement strategy up to 1990 and forecasts to the year 2000 were completed in 1973. Then, in 1975, came the publication of the "General Scheme"[19] for implementing the projects and directing the planning. It sets out the structural-planning principles for setting up and developing the group settlement system at three levels: local, regional, and national.

According to the General Scheme, settlement development is meant to create the urban conditions for man's comprehensive development and to lay the foundation for expansion and rational location of the productive forces in order to strengthen the factors of intensification in social production by making full use of scientific and technological advances. These aims can be furthered by arranging settlement to support the concentration of production, by locating industries more evenly throughout the country, and by preserving valuable farm land. This is to be done by establishing various settlement systems linked in a specific way with large economic complexes. The group system also matches the main trends in the spatial organization of social life and of industry in the period of the scientific-technological revolution, i.e., the way industry is drawn to the centers of skilled labor and scientific institutions; its concentration; the formation of the so-called combine-type industry, with territorial divisions of the production process (a center and branch departments); the trends in locating industry, that is, no longer oriented toward single cities but on urban systems; and finally, the growing mobility of people and their stronger claims to choice of workplace and to improving their skills and choosing their leisure activities.

The Soviet writers define the group settlement system as "...rational spatial organization of settlement groups, either urban or rural, marked by integration of housing, work, and leisure and common enjoyment of the surrounding areas." As can be seen, this is a normative definition that is derived, however, from actual processes. Kaplan, Kochetkov, and Listengurt state this in pointing out that "the group systems constitute a new stage in settlement development, but they arise not from elimination of the earlier stages in urban development but with regard for their reality. This is seen in the integration of existing cities and settlements into the sociospatial pattern of group settlement, where they become the functionally structural elements in the system."[20]

The evolution of group systems is actually an integration process based on growing division of labor, i.e., of functions, among the settlements within a given territory. Naturally, in such a process the individual towns and cities change their functions and their structural character. They lose some of the functions they performed in isolation, apart from the group system, "handing them over" through division of labor to other settlements within the system they belong to. With labor divided among towns it is natural that some functions are promoted in one and restricted in another; this characterizes the specialization inherent in division of labor. A necessary element is the growth of economic, social, cultural, and informational interaction among the towns within a system. And we can stress here that an integral part, i.e., both a condition and a result of group system formation, must be a substantial extension of all communications channels among the settlements. What formerly happened within a single town now takes place among the elements of the group system as a whole. In other words, what occurred within the confines of an isolated town, insofar as such towns ever existed, now takes place in a larger space composed of cities, towns, settlements, networks, communications, and so on. To paraphrase, we could say that "settlement becomes city."

The main technical driving force behind the group system stems from the trend toward greater spatial differentiation of the production and service functions of cities and settlements. These functions correspond to the growing social division of labor, which is accelerating in the course of the scientific-technological revolution. The changes induced by that revolution make a steady advance in industrial specialization, and certainly specialization of other activities as well, an essential condition for efficiency. Projected into the concept of socio-spatial organization, this means that the fully developed group settlement system will encompass a network of specialized complexes: automated complexes of industry, agriculture, transportation, scientific and education complexes, those for rest and free time, complexes for health, sports, recreation, and culture, and so on.

The authors of the theory emphasize that differentiation of the economic bases and division of labor among settlements cannot in itself guarantee the formation of an integrated group system. Differentiation and division of labor merely provide the groundwork for another very important process, namely, the integration of urban functions, in which the forces determining group system formation make themselves most clearly felt.

The group system concept throws new light on the problems of city growth, the stagnation of small towns, and agglomeration. One can say that the system to some extent replaces the large and largest urban formations, retaining their advantages and dispensing with their shortcomings. This enables small towns to exist so long as they become parts of the system (in the Central Economic Region of the Russian Federation, fifty small towns, i.e., with fewer than 50,000 inhabitants, have become parts of systems) and makes it possible to have a more efficient arrangement of industrial urban agglomerations which, in some cases, are rather like spontaneous precursors of the group settlement system. Moreover, since group principles can be applied on a small scale by forming settlement systems around small towns or farming communities, they also provide an instrument for eliminating differences between town and country.

The General Scheme suggests that the group systems to be established in the USSR have three categories according to the size and economic potential of the core cities, i.e., large, medium, and small systems. They should complement each other and embrace the entire settled territory of the Soviet Union. In this sense the concept is not only a methodological principle for the whole state, it also provides practical guidance for settlement policies over the entire inhabited area. This is really the first time that principles of territorial development have been formulated in the USSR on such a large scale.

The <u>large group systems</u> are to be based, according to the General Scheme, on existing urban agglomerations, and they are to follow the zonal development type, i.e., with priority for the outer zones. This will slow down growth in the major central cities of the group systems, where a reduction is envisaged in the absolute rate of population increase compared with present trends.

To achieve slower growth of the largest cities — those with over one million inhabitants — the scheme recommends forming large group systems as counterweights to the cities but located within the same economic areas. This would provide a kind of parallel relief system.

The General Scheme also envisages small and medium group systems progressing up to 1990 mainly through growth of their central cities and improvement in their transportation links with other towns and settlements. This is termed the <u>central type of development</u>.

To achieve greater regional balance in the urbanization level and to enable the principle which, in common with the Polish regional planners, we can term polycentric concentration to be applied, the scheme recommends that a series of large regional centers for group settlement systems be created. They are especially necessary in Siberia and the Far East.

The concept also states the intention to develop the settlement network over the entire country in the period 1976-90. The main objectives can be summarized as follows: selecting the cities that will function as regional centers and assessing forty potential regions for group settlement development; accelerating the growth of cities in the eastern parts of the USSR; slowing down more decisively the growth of cities with more than 500,000 inhabitants and of agglomerations in more developed regions, and basing group systems on them; expanding the network of cities with more than 100,000 to 500,000 people, the system centers, while actively developing medium-sized and small towns with up to 100,000 inhabitants that have good prospects for growth; enlarging villages with development potential and converting them to urban-type settlements; gradually reducing the building of new settlements and, to some extent, of new towns as well.

Regional prognoses have recently become an important element in the work on the general theory and the principles of planning the group settlement system. In many cases detailed studies of the big agglomerations and conurbations have enabled important parts of the general theory to be formulated. This is especially true of the research project at the Moscow Central Institute for Town Building (TsNIIP gradostroitel'stva), where, under the direction of F. M. Listengurt and N. I. Naimark, a study was made entitled "Scheme for Transforming the Settlement Network in the Central Economic Region, with Determination of the Settlement System in the Zones of Influence of Moscow and Other Cities." Between 1926 and 1970 the area with developed forms of the group system expanded from 3 to 20% of the region's total area, and 65% of the population lived in it. Forecasting projects that the areas covered by these settlement forms will increase to 50-60% of the region, and that they will include 90% of the population.

G. N. Fomin identified group systems in the Ukraine in a similar way,[21] and the theory was then applied in a number of prognostic regional studies for the Urals, some parts of Siberia, and other regions.

Some of the problems connected with applying and implementing the group settlement concept are now under discussion. A serious question for theory and practice is how many towns and settlements can be regarded as already belonging to an integrated system. Depending on the accessibility criteria used, the areas and populations of group systems can be interpreted more or less widely. B. S. Khorev and S. Smidovich[22] point out, for instance, that in 1976, 40% of the towns in the Russian Federation fell in the category of "agglomerated settlements," i.e., those lying fewer than 60 kilometers from higher centers. The remaining towns, which can be termed "autonomous," lay outside the group settlement systems. That means, of course, that even with the growing importance of the group systems, attention must also be paid to developing these places as well.

Some town planners point out that despite all their advantages, the group systems can cause certain diseconomies. Should the division of functions among the centers, especially in the large regional systems, be carried too far, transportation costs could become heavy. It seems, then, that deciding the degree of polyfunctionalism for settlements in an integrated system will be among the most difficult problems in theory and practice for the future of the group systems.

Implementing this strategy will also require a change in approaches to economic and town planning.[23] O. S. Pchelintsev has pointed out in a recently published paper[24] that in tackling urban problems, it will be necessary to develop suburban areas more intensively, to link cities with their surroundings, to stimulate decentralization of industry and population in urban areas, to establish an integrated system for utilizing urban and suburban land, to concentrate investment and urban infrastructures in the hands of town administrations, to develop special forms for suburban housing construction, to establish protective green belts around the central towns of the group systems, and also to create such systems both in the "old" and the newly settled regions, linking this to decentralization of science, culture, and administration by setting up systems of regional scientific-technological centers. The idea of the group system in this regional scientific-technological center form is also, according to Pchelintsev, an important instrument for future development in Siberia.

The emphasis on the evolutionary approach to settlement research that recently marks the Soviet studies inevitably led some writers to consider the future, especially of cities and their agglomerations. "What will come 'after the cities,' after this form of settlement has exhausted its potential for man's development and the advance of production? The question is not premature. At the present time a number of economically advanced countries, both socialist and capitalist, are experiencing to the full the processes of forming 'postcity types of settlement,' a kind of transurbanization."[25] The new settlement forms, the first symptoms of which are emerging, differ from the current types not only in quantity but also in quality; they are a kind of antithesis to the city, but not a return to the small or medium-sized town. According to a quite numerous group of writers, development is tending toward the formation of large urbanized regions, the basic mechanism here being, according to Pchelintsev, a process of suburbanization, i.e., continuing decentralization and spreading of urban industries and population to more distant parts of surrounding areas. However, this forecast, too, is based on certain assumptions that will need further verification; among other things, it depends on how far other forces (e.g., the need to save energy throughout the transportation system, restrictions on taking over farm land, etc.) are working to counter the decentralization process, making for more compact city growth. It can be assumed that the experts will very soon be discussing the effects of all these processes, and that the group settlement concept will be given more precise shape. The great importance of the problems connected with the country's regional policy and its settlement development, an appreciation of the economic significance attaching to rational organization of settlement, together with a broad exchange of views on these matters over the past decades can leave no one in doubt that management of urbanization will continue to be among the foremost subjects for research and scholarly interest in the Soviet Union.

3

POLAND

Before World War II Poland, as a predominantly agrarian country in which agriculture accounted for 45% of the national income, was among the weakly urbanized parts of Europe. In 1927-31 there were 568 towns. The number increased in the immediate prewar years to 611, inhabited by 29% of the population. In the first postwar years, despite the great loss of life during the war and the extensive destruction of towns, the urban population was much the same. Dangel[1] states that in 1946 roughly 7.5 million persons out of a population of 24 million were living in towns, i.e., 31% of the total.

The first years of economic reconstruction and the beginning of socialist industrialization were, however, linked with fairly rapid concentration of population in towns, so that by 1950 the proportion of urban population had reached 39%. The rate increased after 1950, reaching its peak in the decade 1950-60. In that period the urban population increased by 5,157,000, and the percentage share reached 48% in 1960. The main factor here was the concentrating of industrial investment and housing construction in cities, primarily the county capitals and the industrial cities of Silesia. After 1960 the rate decreased, as can be seen from the fact that in the decade 1960-70, the urban population rose by no more than 2.6 million. There was also a new phenomenon in the countryside, namely, an absolute increase in rural population for the first time since the war, the decline having been continuous between 1945 and 1960. The slowing down of concentration was due to a number of circumstances, primarily the overall decline in natural population increase, but also the fact that a relatively high level of urbanization had already been achieved and that housing construction was going more slowly, while rural communities on the peripheries of urban regions had become more accessible.

The postwar period, with its rapid urbanization, was also a time of great industrial development, of steadily increasing demand for labor, "which was obtained, when local resources were exhausted, from reserves in the less economically activized localities."[2] This led to fairly extensive migration, which accounted for roughly a quarter of the growth recorded by Polish towns and cities at the height of their expansion between 1950-60.

In the sixties, when more than half the population was living in urban communities, the Polish People's Republic also became one of the "lands of cities."[3] By 1970 there were twenty-four cities with more than 100,000 inhabitants, and around half the urban population and nearly a quarter of the country's total population lived in them. The period 1950-70 saw most rapid growth by cities with more than 200,000 inhabitants. The majority were the core cities of industrial urban agglomerations, the largest being comparable in size and population to similar forms of urban settlement in the Soviet Union. The data on the number of people living in agglomerations differ according to the methods used to delimit them, and the differences can be quite pronounced. Since the procedure used by Poland's Main Statistical Office (GUS) agrees most closely with those used by international statistical organizations, we shall first give the 1965 figures from that source. The GUS definition is based on the principle of so-called metropolitan areas and is a definition of agglomeration in the narrower sense. Agglomeration in the broader sense is used by Malisz, and especially by Leszczycki, Eberhardt, and Heřman. We shall confine ourselves to agglomerations having, according to the Leszczycki et al. definition, populations of more than half a million in 1966. The GUS figures are given first, followed by Leszczycki et al.

Table 13

Urban and Rural Population in Poland, 1945-70

Settlement	1945		1950		1960		1970		Changes in index values		
	abs.	%	abs.	%	abs.	%	abs.	%	1950-60	1960-70	1945-70
Poland, total	23,895	100.0	23,035	100.0	29,795	100.0	32,605	100.0	119.0	109.4	136.5
Urban, total	7,602	31.8	9,244	36.9	14,401	48.3	17,031	52.2	156.0	119.6	224.1
Rural communities, total	16,293	68.2	15,791	63.1	15,394	51.7	15,574	47.8	97.5	101.2	95.7

Source: A. Stasiak, "Specyfika polskiej drogi urbanizacji w świetle wyników NSP, Miasto, 1971, no. 6, p. 6.

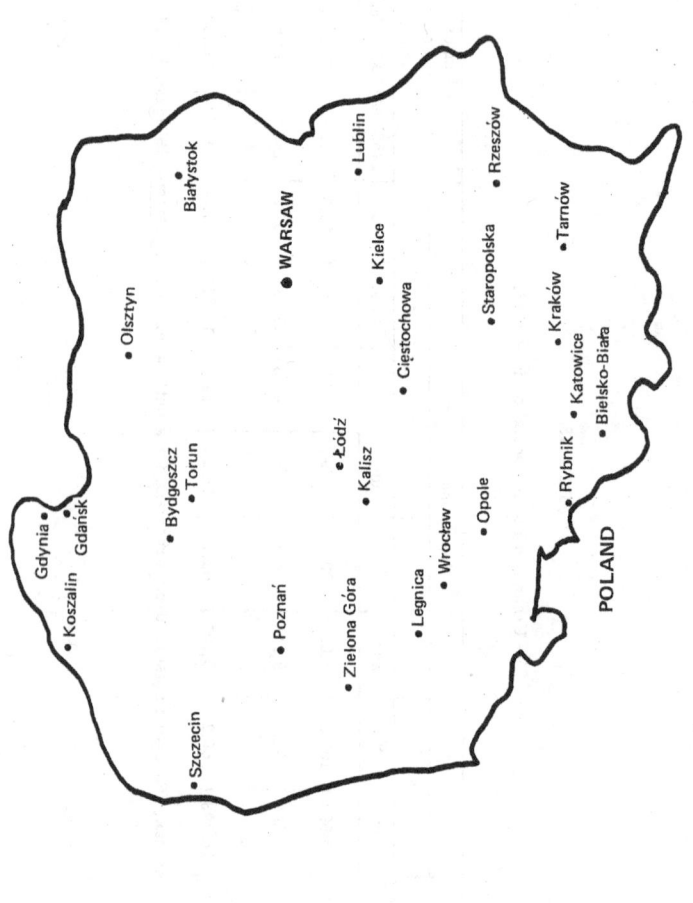

Poland

The largest Polish agglomeration is in the Upper Silesian Basin, with Katowice as its core. The population in 1965-66 was 1,953,000 in the narrower, 3,032,000 in the broader delimitation. Second was the Warsaw agglomeration, with 1,497,000 — 1,782,000 inhabitants; then came the agglomerations of Łódź, with 744,900 — 936,300; Gdańsk, with 595,800 — 618,000; Kraków, with 537,000 — 985,000; Wrocław, with 474,200 — 510,000; Poznań, with 464,500 — 501,000; and Bielsko-Biała, with 485,000 — 553,000.

Although the spatial distribution of Polish cities that have regional functions is fairly symmetrical, the degree to which the urban populations and settlements are concentrated in different regions varies considerably. The majority of the urban agglomerations lie in the southern and central parts of Poland, where a belt of so-called urbanized territory also runs (according to the Leszczycki definition).

The cities in other regions are smaller, and their network is thinner. This is especially true for the northeastern parts of the country, which are less settled and also less urbanized. Western sections are similarly more sparsely settled, although their urbanization level is higher than the national average. The unevenness in settlement structure and concentration of population in Poland has been pinpointed by R. Grabowiecki: "About 33% of the population inhabit the southern provinces — Rzeszów, Kraków, Katowice, Opole, and Wrocław, which cover about 24% of the country's territory. About 70% of the urban population of Poland and about 36% of the total population live in seventeen urban agglomerations; among them, 4.5 million people, i.e., about 24% of Poland's urban population, live in the two biggest agglomerations — Katowice and Warsaw. Decisive for the role of these two agglomerations is the fact that about one quarter of all industrial job opportunities are concentrated in their territories (1970). Also living in the Warsaw agglomeration are about 60% of people engaged in science, culture, and art, and the Katowice agglomeration accommodates about one fifth of the capital assets of industry."[4]

The present settlement structure and the macroregional

variations in the levels and forms of urbanization have undoubtedly been considerably influenced by the distribution of raw materials, but also significant have been factors of historical nature. Goldzamt writes that Poland and Czechoslovakia lie on the borders between two settlement structures, namely, the dense, at times very dispersed settlement of western and central Europe, and the medium to sparse settlement of eastern and northern Europe. "Southwestern Poland, together with the Czech basin and the greater part of Moravia, has a settlement structure similar to the structures of areas in central and western Europe, such as the southern parts of Germany, Austria, northern Italy, etc. Eastern Poland and Slovakia are in a settlement zone that is transitional to the eastern European type. Belonging to this transitional zone are the western parts of Byelorussia, a large part of the Ukraine, Transylvania in Romania, the northern settlement belt in Hungary from the Czechoslovak frontier to the Great Hungarian Plain, as well as the eastern part of Croatia."[5]

Table 14

Proportions of Polish Population in
Rural Communities, Towns, and Urban Settlements
by size, 1950-70

Sizes of urban communities	Population in urban communities, in %		Changes
	1950	1970	
to 5,000	4.3	3.4	-0.9
5,000-10,000	4.5	4.8	+0.3
10,000-20,000	4.2	6.8	+2.6
20,000-50,000	6.2	9.0	+2.8
50,000-100,000	3.4	5.6	+2.2
100,000-200,000	6.7	6.7	0.0
200,000 and over	9.7	15.9	6.2
Urban total	39.0	52.2	+13.2
Rural total	61.0	47.8	-13.2

Source: A. Stasiak, "Specyfika polskiej drogi urbanizacji w świetle wyników NSP," Miasto, 1971, no. 6, p. 9.

Poland

Poland's postwar urbanization problems were also affected to a considerable extent by macroregional differences in economic and social conditions. These stemmed from the historical partition of the country into Prussian, Russian, and Austrian segments. The economically most advanced area was the western, Prussian part. As B. Malisz records, in the period 1860-1913 these western territories, which included Upper and Lower Silesia, underwent rapid economic development, growth of cities, and the spread of a relatively dense urban settlement network. The territory belonging to tsarist Russia also began to be industrialized before World War I, although not as rapidly as did the Prussian segment. Least economically advanced was the area annexed by Austria, where before World War I only six people in a thousand were employed in industry, whereas in the Prussian segment the figure was sixty-three. Despite some changes in the interwar years, when, for instance, industrialization started in parts of central Poland and communication links were forged between the three sections of the country, which were also administratively integrated, the legacy of traditional regional differences and contrasts remained an obstacle to be overcome by postwar regional policies; its effects are felt to this day.

Some Polish writers therefore distinguish between "strong" and "weak" regions; in describing the weaker ones, they refer to the "Polish mezzogiorno."[6] According to their analysis there was a leveling in some respects between 1950 and 1970, especially in the employment structure, i.e., as regards differences in the proportion employed in industry and the demographic structure. On the other hand, there are still considerable differences between the strong and the weak areas from the standpoint of national income produced. This is due to a development strategy that emphasizes capital-intensive investment in the strong regions and labor-intensive investment in the weak ones.

From the standpoint of our central theme in this study, it is important that these disproportions have also persisted due to delayed investment in the infrastructure and also to the relative lag in the process of urbanization behind industrialization.[7]

Urbanization in Socialist Countries

In connection with uneven urbanization in their country, Polish writers point to the sociologically important circumstance that towns in the areas with well-developed and old settlement networks have quite good housing stock, good infrastructures, and fairly developed tertiary sectors. They refer to these regions as "old urbanization areas," in contrast to areas with uneven and undeveloped urban networks and usually with inferior infrastructures. These are the "new urbanization areas" established by the planned industrialization of formerly backward regions with large reserves of manpower.

Another factor of sociological significance is the so-called urbanization of the Polish countryside, which means that there has been a long-term increase in the proportion of people living in the country but not working in agriculture. In certain periods, e.g., in 1960-70, the increase in nonagricultural rural population even exceeded the increment in urban population. The crude figures on the absolute increases in nonagricultural rural population are also interesting. Whereas in 1950 they numbered 4 million persons, the census figure for 1960 was 4.7 million, and by 1970 the total had risen to 6.7 million. The process is likely to continue, changing current trends in Polish urbanization, which has been marked by rapid concentration of population in the major urban centers. In any case, it has already created today a strong category of mixed industrial-rural population that in many respects has some features of the urban life style but also maintains numerous economic ties with the traditional rural community.

Urbanization and Settlement Strategies

The Polish People's Republic is among the countries that started work on regional and settlement policies for developing its whole territory during the first postwar years. The work followed the progressive traditions of the prewar days when, along with theoretical studies, of which the most important was Warszawa Funkcjonalna [Functional Warsaw] by J. Chmielewski and Sz. Syrkus (published in 1934), the late thirties saw the founding of a state institute to work out a nationwide regional plan for Poland.[8]

Shortly after the end of the war, in 1946, a decree on physical planning was issued that distinguished three planning levels: (1) the plan for the whole country; (2) regional plans; (3) town plans. Under this legislation the Main Office for Physical Planning was set up under the Ministry for Reconstruction, and the first strategy for the whole country was worked out. This strategy was published in 1947, followed in the same year by an atlas documenting its ideas. It is typical for Poland that the 1946-47 concept was already described as a "directional structure," an idea referring to the network of main transportation lines and of settlements linked by communications. This was, in fact, one of the first versions of the nodal belt concept, which is still being extended and perfected in Poland. The plan envisaged three stages: the first, to 1950, was aimed at reconstruction; the second, to 1965, was to carry out complete industrialization; and the third, to 1980, was to finish total urbanization. Toward the end of the third stage, when a population of 32 million was anticipated, 20 million people were to be living in towns. Although demographic growth was somewhat underestimated, the figure for urban population in 1980 was remarkably accurate, matching the level that has now been reached.

The plan proposed to decentralize industry and to form three new industrial areas in the less advanced regions. However, work was halted in 1950 owing to conflicts that arose between the organizations responsible for physical planning and those concerned with economic planning. In the book already cited, B. Malisz writes of this period: "Looking with the hindsight of a quarter of a century at this work in its entirety, one must say objectively that it was a time when the ideas of physical planning blossomed in Poland, when currents of thought formed that have remained relevant to this day. A weakness was the poor linkage of physical planning with considerations of economic development, especially with planning of the national economy, which had started to advance at that time." The reasons for the conflicts were complicated and numerous. An important part was played by the existence of an institutional division of responsibility for the two types of planning. In the second half of the fifties the first steps were taken in Poland toward long-term planning, i.e., toward laying the main lines for economic development covering the coming fifteen to twenty years. This was the time span applied in the national plan and in the plans for individual regions. At the same time, criticism arose about measures to separate the three levels of physical planning that had been introduced in 1950. These and other factors got the nationwide plan started again. A milestone was the promulgation of a law on physical planning in 1961. This legislation stressed the link between planning the national economy and the long-term plan for the country's regional and settlement development. The favorable conditions provided by the legislation led, in 1967, to the setting up of an independent "Working Party for the National Plan" attached to the Planning Commission. At the same time, planning theory was developed, thanks in particular to the interdisciplinary Commission for the National Development Plan under the auspices of the Polish Academy of Sciences (formed in 1958), which drew on the country's valuable traditions in the fields of regional economics and the location of productive forces (see the works by K. Secomski, T. Mrzygłód, and S. M. Zawadski), urban planning (see the papers by J. Chmielewski, B. Malisz, and J. Goryński), geography (K. Dziewoński and S. Leszczycki), and also sociology, which in the sixties started intensive studies of urbanization, as evidenced by the work of J. Ziółkowski, S. Nowakowski, P. Rybicki, and Z. Pióro. Also helpful in working out settlement concepts was the existence of various research institutes, well-developed division of labor in research, and a lively awareness that urban development and the organization of settlement were a vital part of the country's problems.[9]

The second generation of forecasts and settlement strategies, published in 1967-68, came from this environment. Differences of opinion arose most frequently, both during the drafting and after publication, around the question of the degree to which settlement should be concentrated or dispersed. Essentially the argument reflected the search for the best balance between economic efficiency and the social goals of a socialist country. It was further stimulated by the existence of settlement structures inherited from the past, with the great Upper Silesian industrial agglomeration, on the one hand, built on its raw material base, and on the other, with the less developed regions of the northeast, as well as Kielce Province and other areas. Other questions followed more or less directly from this — should development continue along current lines, i.e., by concentrating social activities and people in the major agglomerations, or should the policy be to restrict the growth of large formations and to develop and activize large numbers of small and medium-sized towns? What are the economic and social advantages offered by cities of different sizes? Connected to some degree with the central questions was a further problem — finding a suitable method for gradually eliminating the social differences among the regions. A solution to the controversy between concentration and dispersal was already being sought in the sixties in the shape of new forms of settlement that would overcome, or at least moderate, the disadvantages of both trends. The idea of belt settlement development (Malisz) emerged or, more precisely, was revived, and the ideas for urban regions were formulated (Goryński) that are so popular today in settlement strategies.

The two forecasts produced in the second half of the sixties were not in fact directly opposed alternatives because both of them rejected both radical dispersal and radical concentration. One, however, obviously tended more toward deconcentration of industry and manpower, while the other inclined more toward concentrating economic and human resources. The forecasts, to the years 1985 and 2000 respectively, were the work of two different institutes. At the National Development Plan Institute, under the leadership of T. Mrzygłód,[10] a strategy was worked out that was meant to keep the population in existing settlements and minimize migration from one province to another; it proposed that industry be developed primarily in selected subregional centers. While the national physical plan for 1966-85 drafted by Mrzygłód's group relies primarily on basic principles for the economic development of the whole country, it is also concerned with distributing the various elements in such a way as to contribute to eliminating the major inequalities in living standards among the different regions. That is to say, in some areas, e.g., Białystock, Kielce, Rzeszów, Lublin, and Olsztyn provinces, living standards would rise more rapidly than elsewhere. The strategy includes increasing job opportunities in these areas while reducing the rate of increase of openings in Warsaw and Katowice provinces. Mrzygłód rejects the extremes of concentration or dispersal of investment, the first mainly because per capita costs for housing and civil construction and for technical infrastructure are higher both in terms of investment and operation in cities and agglomerations than in other urban centers. All-out dispersal would, however, "result in a very serious reduction of investment efficiency" because the advantages of having public utilities and other infrastructure components in common would be lacking. Infrastructures would have to be constructed or extended in a whole series of towns and urban-type settlements.

Its authors therefore describe this as a "third" variant, that is, a middle course meant to concentrate new industrial plants in several dozen centers where both intensive and moderate industrialization would take place. There should be sixty-seven centers for intensive industrialization and twenty-two for moderate industrialization. The strategy assumed that it would be necessary to carry out active and passive deglomeration in areas where industry had become overconcentrated. These deglomeration areas include Upper Silesia, Warsaw, Łódź, Kraków, Poznań, and Gdańsk-Gdynia. The towns selected for intensive and moderate industrialization are fairly evenly distributed over the various provinces. The proposal implies reversing existing urbanization trends in Poland whereby population had been predominantly concentrated in nine major agglomerations. Implementing Mrzygłód's concept would mean that population in the deglomeration areas would increase by 1985 by 1.3 million, i.e., 124%; in the intensive urbanization towns, by 3.4 million, i.e., an increase of 240% compared with 1966. In the other towns the figure would be lower, but not as low as in the deglomeration areas. The growth would come, apart from natural increase in the urban population, from rural migration. It is envisaged that the period 1966-85 should see an increase in the urban population of 9 million, to reach a total of 24.7 million. The rural population would drop by about one million. Urbanization would advance, however, primarily in the areas where it has hitherto been weak and where the urban element does not exceed 30% of the population.

The second strategy, drafted by the Institute of Town Planning and Architecture under B. Malisz,[11] envisaged fairly substantial, albeit regulated growth of the biggest agglomerations. It was projected to the year 2000, with a phase described for 1985. The group that submitted the concept anticipated the development of so-called urban groupings and urbanized belts linked to nodal urban areas. In general it is evident that the proposal devotes considerable attention to the spatial arrangement of the settlement network. It also envisages the future settlement system as being hierarchical, with six levels: capital city, second-level centers (large industrial and port agglomerations) and centers of wider than regional significance, regional (provincial) cities, subarea centers, and finally, district and other small towns. It is evident from the model that Poland's future settlement structure will be shaped by five factors: future location of productive forces, inertia of existing settlement, location of nonproductive activities, technological advance in transportation, and costs connected with different forms of population distribution. According to Malisz, the distribution of natural resources and productive forces influences the settlement network because it creates belts stretching between extraction areas and outlet centers. In contrast, the tertiary sector — services in the widest sense — is noted for its spontaneous tendency to concentrate and to form systems of centers. In combination the two forces then give rise to the nodal belt system of settlement that Malisz had already summed up in his book Outline of a Theory of the Formation of Settlement Systems,[12] published in 1966. As we shall see later, the nodal belt system of settlement is one of the foremost hypotheses in the entire Polish settlement theory.

Malisz based his proposal on an optimistic version of Poland's demographic development that assumed that by 1985 the population would number 39.6 million, and by 2000 about 45 million. Urban population should rise from 15.7 to 24.6 million by 1985 and to 32 million by the year 2000. That would be a doubling of the urban population. Realistic critics, correctly taking into account housing construction potentials, have pointed out that such rapid concentration of people in the towns must cope with the capacity to build housing for them. Moreover, later demographic forecasts have shown that population growth will not be at the rate assumed earlier, and more recent settlement forecasts anticipate a population of 38-40 million by the year 2000 rather than 45 million.

The third generation of settlement strategies originated in the seventies in response to several new circumstances in the realm of ideas and in the institutional and organizational sphere. In the first place, it is evident that during the seventies the national physical plan for Poland began to gain in importance and to acquire a certain autonomy vis-à-vis economic planning. Although its formulation was based on the general premises of long-term economic planning, it was not a passive projection of them into the physical dimension. At the same time, the social goals embodied in the national physical plan began to influence the economic plans. The institutional base for working on the plan was also reinforced, and a series of questions concerning the country's physical development was accorded a principal place in national research programs. This yielded a set of research papers on "The Foundations of the Country's Physical Development," with the work of several dozen research institutes being coordinated by the Polish

Poland

Academy of Sciences. From this project and from work by the academy's forecasting commission "Poland — 2000"[13] came the third generation of settlement and urbanization strategies.

The strategies programmed development to the year 2000, and the State Planning Commission then worked out the "planning" version, which had a shorter time span, dealing with development to 1985. The point is that, on the one hand, there are forecasts that differ according to the degree of concentration they assume for urban population, and that deal with the spatial organization of urbanized zones to the year 2000; on the other hand, there are shorter-term concepts worked out and accepted by government bodies. The concept for settlement to 1985 forms the basis for deciding about locating investment, and it therefore has very concrete implications. In it maximum use is also made of results from the work coordinated by the Academy of Sciences. We shall first turn to this official plan for the country's development and then take note of some of the main theoretical concepts.

The national physical plan originated at the instigation of the Polish government, which in 1971 posed the task of drafting a perspective plan for development to 1990 and with it, as its integral component, a national physical plan. By 1974 a first draft was available in the form of a document from the Planning Commission attached to the Council of Ministers. One of the authors stated the plan's aims as follows:

1. Rational use of space to dynamize the country's socioeconomic development, more effective exchange of goods and services between the centers of settlement and the areas of agricultural production, concentration of industry.

2. Comprehensive development of the population in all areas to try to eliminate excessive differences in living standards and to make centers of civil amenities and the cultural and scientific centers more accessible.

3. Create conditions for the use and protection of the natural environment with regard for the needs of present and future generations.[14]

To achieve these aims, which try to strike a balance among

Urbanization in Socialist Countries

Table 15

Urban Agglomerations in Poland*

Location	Urban population, 1970, in thousands (rounded)	Indicator of increment $\frac{1990}{1970}$ in %
A. Established		
1. Katowice (with Rybnik agglomeration)	2,345	115-118
2. Warsaw	1,790	134-145
3. Łódź	1,000	120-125
4. Gdańsk	735	132-143
5. Kraków (with Chrzanów-Oświęcim agglomeration)	905	130-145
6. Wrocław	630	123-139
7. Poznań	555	137-146
8. Szczecin	430	152-187
9. Bydgoszcz-Torun	510	137-146
10. Sudety	285	124
B. In course of formation		
1. Wielkopolska	460	152-163
2. Bielsko-Biała	215	123-134
3. Częstochowa	260	117-136
4. Lublin	275	147-165
5. Opole	250	141-181
6. Białystok	200	153-179
7. Rzeszów	100	200-220
C. Potential		
1. Legnica-Głogów	170	150-180
2. Koszalin	120	170-195
3. Kalisz-Ostrów	130	122-153
4. Olsztyn	95	158-190
5. Tarnów	100	137-167
6. Zielona Góra	69	198-217

Source: R. Grabowiecki, "Plan przestrzenny Polski do roku 1990," <u>Gospodarka planowa</u>, 1974, no. 4, p. 219.

*The figures are for the cores of agglomerations and urbanized belts; they do not include areas still being urbanized.

Poland

Table 16

Hypothetical Changes in Population Distribution and
Urbanization Areas in Poland to 1990

Area*	Population in thousands		Urban population, in %	
	1970	1990	1970	1990
Białystok	1,175.1	1,270	37.2	55
Bydgoszcz	1,915.1	2,240	50.7	63
Gdańsk	1,469.3	1,870	69.6	79
Katowice	3,701.4	4,340	76.8	83
Kielce	1,890.3	2,010	32.6	51
Koszalin	795.8	990	49.6	66
Kraków	2,772.7	3,130	45.2	61
Lublin	1,925.3	2,030	31.0	49
Łódź	2,432.3	2,550	55.9	68
Olsztyn	979.7	1,130	41.1	55
Opole	1,059.1	1,260	42.6	56
Poznań	2,664.5	3,010	50.3	63
Rzeszów	1,758.2	1,950	27.6	48
Szczecin	898.6	1,160	66.6	76
Warsaw	3,833.3	4,450	57.7	72
Wrocław	2,502.4	2,940	64.9	75
Zielona Góra	884.4	1,070	54.3	68
Poland, total	32,657.5	37,400	52.3	66

Source: R. Grabowiecki, op. cit., p. 217.
*Area here corresponds to the old wojewodstwa and their statutory capitals.

economic, social, and ecological considerations, three principles for spatial strategy were chosen: polycentrism, moderate concentration, and a shift of accent from urbanization of the south to urbanization of the north and east and to strengthening the economies of the central areas. To some extent this is a strategy in opposition to the trend that has operated so far, which has been toward increasing concentration of industry and population in the southern zone.

The authors of the plan emphasize that polycentrism, as they conceive it, means "the opportunity to make fuller use of the economic resources localized in all parts of the country, and it meets the aspirations and needs of the population in all areas." Polycentrism is, then, primarily a socially oriented principle intended to achieve the situation most frequently mentioned among the aims of regional planning, namely, reducing the social inequalities in different parts of the country. However, R. Grabowiecki and S. M. Zawadski also stress that the concept is in no way a return to the discarded "theory of equal distribution of productive forces," because that theory explicitly called for dispersal while the national physical plan is based on concentration. It is a matter of maintaining a reasonable degree of industrial concentration in different localities and areas in order to avoid the urbanization process leading to extreme settlement concentration. The moderate concentration principle can represent, according to its authors, an alternative to the spontaneous growth of big urban agglomerations to be found in capitalist countries.

This principle is manifested in the plan by the emphasis placed on urban agglomerations. They provide an environment favorable to introducing and advancing modern changes in the country, just as the territorial concentration of population, towns, urban-type settlements, and industries, linked with the social and technical-economic infrastructure, is the leading element in the national physical plan.

The strategy distinguishes three categories of agglomeration according to the present size and growth rate of the population. The first includes the ten largest, already existing agglomerations, the biggest, Katowice, having a population estimated by the authors of the plan as 2,345,000, the smallest, Sudety, having 285,000 inhabitants. In the second category there are seven agglomerations in the process of formation, the largest with 460,000 inhabitants (Wielkopolska), and the smallest with 100,000 (Rzeszów). Six potential agglomerations form the third category, the largest with 170,000 inhabitants (Legnica-Głogów), the smallest with 70,000 (Zielona Góra). The urban agglomeration system is supplemented by fifteen urban centers of national significance that perform various service functions for industry. There are also around fifty towns of purely regional significance. The plan also includes about 2,500 rural settlements throughout the country where the administrations of the newly established large communities (gmin) are situated, together with centers serving the rural population and farms; in some cases there are small manufacturing concerns.

The plan also includes quantification of anticipated changes. It calculates that the population in towns of all categories will increase from the 1970 figure of 17.1 million to 24.7 million by 1990, a rise from 52.3% of the total population to 66%. In absolute terms the biggest increment is accorded to the agglomerations (from 11.6 to 16.6 million), but the relative weight of this increment will not be as great as in the other categories of urban settlements. In any case, it is probable that over half the urban population will be living in agglomerations by 1990.

As we have already mentioned, the drafting of the official physical plan was preceded by theoretical forecasts for the development of Poland's settlement system up to the year 2000. They were produced by various people, and they differed in many respects, including in the weight given to concentrating activities, infrastructures, and population in urbanized areas and cities. But it must be emphasized that the differences are not great, as evidenced by the fact that some of the proposals were later merged.[15] First, however, we will consider the ideas that were independently advanced, so that their central themes become clear.

As representative of the views stressing concentration, we can take the study produced under the leadership of Professor Leszczycki's group and the similarly conceived work by R. Karlowicz; a medium degree of concentration is expressed in the proposals by K. Dziewoński and B. Malisz; while P. Zaremba's concept represents an emphasis on smaller concentration.

St. Leszczycki, P. Eberhardt, and St. Heřman[16] proceed from the standpoint that urbanization is a process with historical phases differing in the dimensions and forms in which people and their activities are concentrated. Following the phase of the preindustrial town, lasting more or less to the end of the eighteenth century, came the phase of the concentric industrial town, which marked nineteenth-century urbanization. In the twentieth century we are passing through the agglomeration of urban settlement systems to the phase of their metropolization. This is a result of economic, social, and technological changes which, on the one hand, stimulate the concentration of people and activities and, on the other, make it possible to modify it. Applying their findings on the historically changing phases of urbanization, the authors have constructed a prognosis envisaging 75% of the Polish population living by the year 2000 in sixteen agglomerations, industrial urban centers, and other urban centers. With a total population of 38 million, that would be 29 million people. Of these, 23 million would be in urban industrial agglomerations, i.e., fully 80% of the entire urban population. Of the 6.5 million working in agriculture, 1.5 million would be living within agglomerations and industrial centers. As distinct from the national plan, the Leszczycki group's project envisages more rapid growth of population in urban industrial agglomerations and urbanized territories. It estimates a rise over the 1966-2000 period from 12.3 to 23 million.

Such sharp growth would necessarily lead to changes in the spatial structure of the present agglomerations and to their differentiation. Urban settlement structures will be larger in the future, encompassing an even greater economic and demographic potential than is the case today. Hand in hand with concentration on the national scale there will be deconcentration within the agglomerations in order to ensure good ecological conditions for economic life and for the quality of housing and recreation available to the inhabitants. Naturally the agglomerations will have to occupy more and more Polish territory. In time some will merge, as already indicated by the trends now leading toward merging of the Kraków, Katowice, Bielsko-Biała, Opole, and Wrocław agglomerations (see Figure 2).

The concept for a system of large agglomerations worked out at the Warsaw Polytechnic by R. Karlowicz is a sort of modification of the Leszczycki approach. In this view the large

Poland

Table 17

Forecast for Industrial-Urban Agglomerations
in Poland to the Year 2000,
According to Leszczycki, Eberhardt, and Heřman

Agglomeration	Situation to 1966				Forecast to 2000			
	area in km^2	population in thousands	in %	population per km^2	area in km^2	population in thousands	in %	population per km^2
Katowice	6,124	3,032	9.6	495	6,804	3,600	9.5	530
Warsaw	1,893	1,782	5.6	941	6,633	2,900	7.6	437
Kraków	2,832	985	3.1	348	6,102	1,700	4.5	279
Łódź	587	932	2.9	1,586	3,037	1,500	3.9	494
Podsudety	2,641	722	2.3	273	6,247	1,300	3.4	205
Staropolska	2,691	633	2.0	235	6,967	1,200	3.2	176
Gdańsk	510	618	2.0	1,211	1,792	1,200	3.2	669
Bielsko-Biała	2,564	553	1.7	216	3,093	800	2.1	259
Wrocław	482	510	1.6	1,057	1,908	900	2.4	471
Poznań	455	501	1.6	1,099	1,446	800	2.1	551
Opole	3,114	474	1.5	152	6,224	1,000	2.6	161
Bydgoszcz-Torun	384	399	1.3	1,037	4,750	1,300	3.4	270
Częstochowa	1,521	398	1.3	262	3,894	800	2.1	206
Szczecin	420	331	1.0	786	1,615	900	2.4	557
Lublin	201	230	0.7	1,145	822	400	1.1	486
Białystok	466	172	0.5	307	1,124	350	0.9	311
Total existing agglomerations	26,885	12,272	38.7	456	62,458	20,650	54.4	330
Rzesów-Tarnobrzeg	×	×	×	×	5,234	1,000	2.6	191
Carpathian	×	×	×	×	2,435	500	1.3	206
Legnica-Głogów	×	×	×	×	2,438	500	1.3	206
Kalisz-Ostrów	×	×	×	×	1,418	350	0.9	246
Total agglomerations	×	×	×	×	73,983	23,000	60.5	311
Other parts of country	284,845	19,279	61.3	68	237,747	15,000	39.5	63

agglomerations can even now be regarded as a subsystem
of the overall settlement system in Poland, and their importance will increase. The author believes, however, that
in the future, along with development of the nodal belt
structures which the majority of Polish town and regional

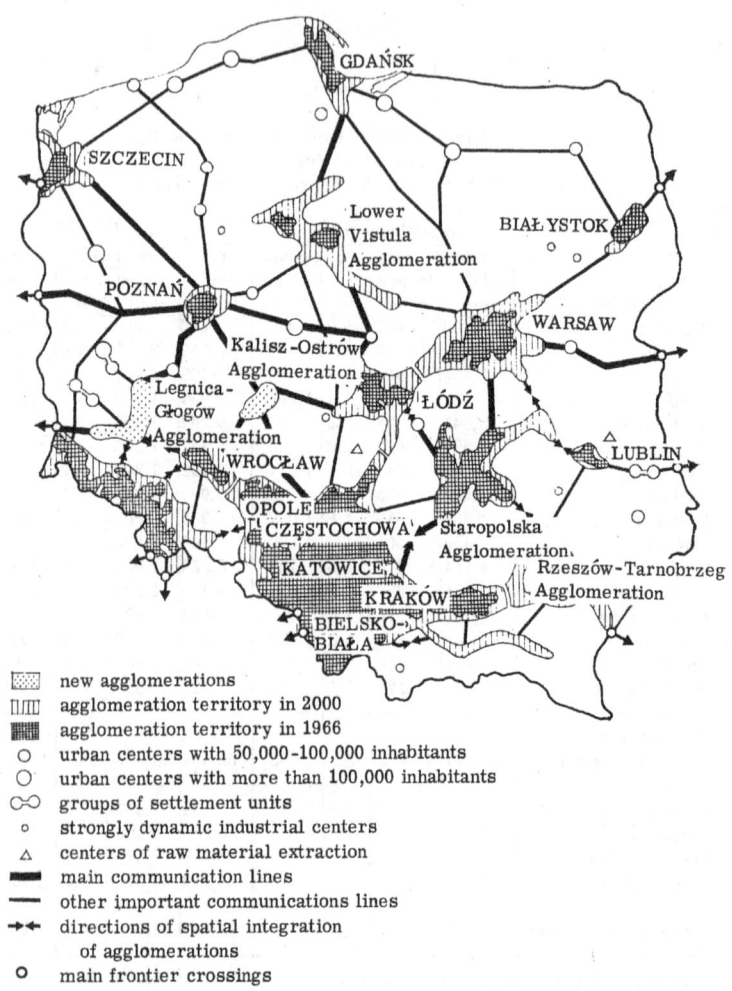

- ▦ new agglomerations
- ⧇ agglomeration territory in 2000
- ▤ agglomeration territory in 1966
- O urban centers with 50,000-100,000 inhabitants
- ◯ urban centers with more than 100,000 inhabitants
- ∞ groups of settlement units
- ∘ strongly dynamic industrial centers
- △ centers of raw material extraction
- ▬ main communication lines
- — other important communications lines
- →← directions of spatial integration of agglomerations
- O main frontier crossings

Figure 2. Concept for urban agglomerations system in Poland, after S. Leszczycki, P. Eberhardt, and S. Herman.

planners favor, "the nodes will gain in importance, while the belts linking them will retreat into the background. We must expect the large agglomerations to establish a system of cooperation that will manage, on the whole, without connecting belts; they will be omitted, as will the small settlements occupying the areas lying between agglomerations."[17] It can therefore be assumed that long belts of settlement will, in particular, be of less importance. The introduction of new means of transportation leads to such a conclusion, and the view is also supported by the development of air transport, which tends to encourage nodes rather than belts.

The prognosis advanced by the prominent Polish geographer K. Dziewoński occupies the middle position. The same standpoint has been adopted in the new proposal by the town planner B. Malisz. Dziewoński assumes[18] that the framework for settlement and urbanization development should consist of thirty urban regions with areas averaging 10,000 km^2. On that area, preferably in the center, there should be at least one city with a minimum population of 150,000 to 200,000. The cores of these urban regions would lie an average 60 kilometers from the peripheries, that is, an hour's journey, which would enable urban groupings to form around them composed of towns and settlements linked to each other by commuting. Professor Dziewoński believes that one third of such urban regions exist de facto in Poland today, one third are in process of formation, and only one third show no signs as yet of emerging. His theory is essentially also a polycentric strategy, which the author links with eliminating regional differences.

In the seventies B. Malisz[19] produced a new version of his model of nodal belt development of settlement in Poland. He assumes that the population will be 40 million by the end of the century, of which 8 million will live in the countryside. The remainder, i.e., 32 million, will have to be accommodated in urban centers, which is twice the number living in towns in 1971. That is no small task, and it requires the assembling of considerable economic resources.

For the prognosis to be realistic it must proceed from existing long-range trends, but it is not obliged merely to adapt passively to them or to extrapolate solely from present trends. In Malisz's view, that is the weakness of the Leszczycki group's work. The nodal belt model also envisages the most rapid growth in the large agglomerations and then in those of the so-called secondary nodes. But since this model emphasizes spatial organization, its implementation would avoid the shortcomings that are problably linked inevitably with the present mode of developing large agglomerations. Settlement is concentrated in the nodal areas and the belts running out from them. This enables urban life to be rationally organized while also preserving space for farming and recreation in the "loops" between belts. The nodes throughout the system would accommodate, in particular, the tertiary sector; industry and residential areas would be in the belts. Malisz maintains that in a model of the settlement network conceived along these lines, the sizes of the population groupings cease to be important. At the turn of the century the whole network will form a single system. Advances in communications will facilitate access to the big centers even from hitherto remote places (over 300 km), and the smaller centers will be accessible from places 50-60 km distant. A further advantage of the model is, according to its authors, that the rural population — once the center concept for servicing it (hexagonal system) is abandoned, and it is concentrated in belts along with the urbanized zones — will be in the same situation as people living in urbanized areas. Moreover this is no utopia, for in 1965, 87% of Poland's urban population was already living in belts as defined by Malisz, and the growth of towns lying outside the belts was half that in those lying inside them.

The last forecast to be dealt with is the attempt by P. Zaremba.[20] Its initial version was most radical in proclaiming support for small towns: "I regard as sensible an urbanization policy that would enable the small towns to be preserved, on condition that they are modernized and receive all the amenities provided by contemporary civilization." The same policy should be adopted for the medium-sized towns, for within a few years the proportion of the urban population living in towns of fewer than 50,000 inhabitants will drop from 45% to around 30%, whereas the proportion in the medium-sized towns (50,000 to 200,000) could rise from the present 24% to 35%. Zaremba criticizes the one-sided economic reasons for supporting large city growth, and he points to the insoluble ecological problems of agglomeration and to its social consequences. He does not believe that deglomeration by means of satellites can radically change the situation. He concludes his thesis by saying that forecasting the spatial structure for Poland's urbanization should envisage about ten main urban agglomerations and should promote the development of medium-sized and small towns lying along the belts of intensive development. And so Zaremba's concept is, in its turn, also a variant of the nodal belt model for settlement development.

Neither the theoretical concepts nor the government plan have yet been able to take into account the administrative reform of 1975, which replaced the three-tier administrative structure with a two-tier system. The first level of administrative units is formed by the gminy, which group several formerly independent communities, and the second are the wojewodstva (provinces). The reform has increased the number of provinces from seventeen to forty-nine. Some theoreticians, for instance, Malisz,[21] have expressed the view in studies published since 1975 that the reform is essentially in line with the national plan for physical development. It seems, however, that one of its side effects may be — in our judgment — that the concentration process will be slowed. The reform may strengthen the polycentric and deconcentration element, but it will probably not affect the macroregional goals of physical development, which are meant to abolish the traditional division of Poland into vertical belts — those further to the east being the less advanced — and to substitute a horizontal arrangement enabling the country's natural conditions to be used to the fullest.

4

THE GERMAN DEMOCRATIC REPUBLIC

The German Democratic Republic is the most highly urbanized of the socialist countries; in 1975 about 75% of the population lived in urban communities and around 80% in urban regions. The development of settlement during the period of socialist construction was based on the old system that had grown up under feudalism and been reshaped by capitalist industrialization. It bears all the marks of central European settlement structures, especially in the south of the country, which includes the old European industrialized territory of Saxony, dating from the second half of the nineteenth century. Development continued in the first half of the twentieth century, when rapid urbanization took place on the territory of the present GDR, accompanied by fairly large increases in population and by a process of polarization.

In 1871 Berlin already had a population of nearly one million; but on the other hand, many agricultural areas, particularly in the north and northeast, remained static, with the settlement system changing only gradually. A feature of the urbanization process toward the end of the nineteenth century and in the first decades of the twentieth was the rise of large industrial agglomerations and the formation of a relatively stable hierarchy of large towns. Consequently the GDR inherited a well-developed and stabilized settlement system. The effect can be seen, among other things, in the fact that concentration of population is going more slowly than in the other socialist countries.

The German Democratic Republic

Owing to the high degree of industrialization and urbanization existing at the time of its foundation and the slight population increase, the GDR has not experienced so radical a transformation of its settlement structure as has been the case with the countries undergoing rapid industrialization and at the same time registering relatively fast demographic growth. The basic elements of the settlement structure have been stabilized for some time, and economic development occurs within the framework of this established structure.[1] But that is not to say that there has been no change or that no new elements have emerged. As in the other socialist countries, the GDR has applied the principle of stimulating the development of its industrially backward areas; in the northern and eastern regions, a number of new industrial centers have been established, and industries have been expanded in existing towns. In Rostock the port was enlarged, a shipyard was opened, and the fishing industry developed; Schwedt, where the oil pipeline from the USSR ends, has become a petrochemicals center; metallurgy is developing in Eisenhüttenstadt, and electronics in Frankfurt-on-Oder. Another new industrial center is developing in Schwerin.

Postwar economic and urban reconstruction has led, as in Poland, to investment flowing to a few of the major industrial and administrative centers. Legislation enacted in 1950 included as part of the national economic plan the reconstruction of Berlin, Dresden, Leipzig, Magdeburg, Karl-Marx-Stadt, Dessau, Rostock, and a number of other cities. This tendency to concentrate investment in the major cities is a permanent feature of GDR settlement policy. Despite this, however, the populations of these places have increased very gradually over the past twenty-five years or have remained almost stable, the exceptions being Rostock and the new 100,000-strong town of Gera. The strongest growth has, in fact, been recorded by towns of fewer than 100,000 inhabitants, thereby reinforcing a traditional feature in the country's settlement structure, i.e., concentration of the urban population in centers with populations of 20,000 to 50,000[2] and 10,000 to 20,000.

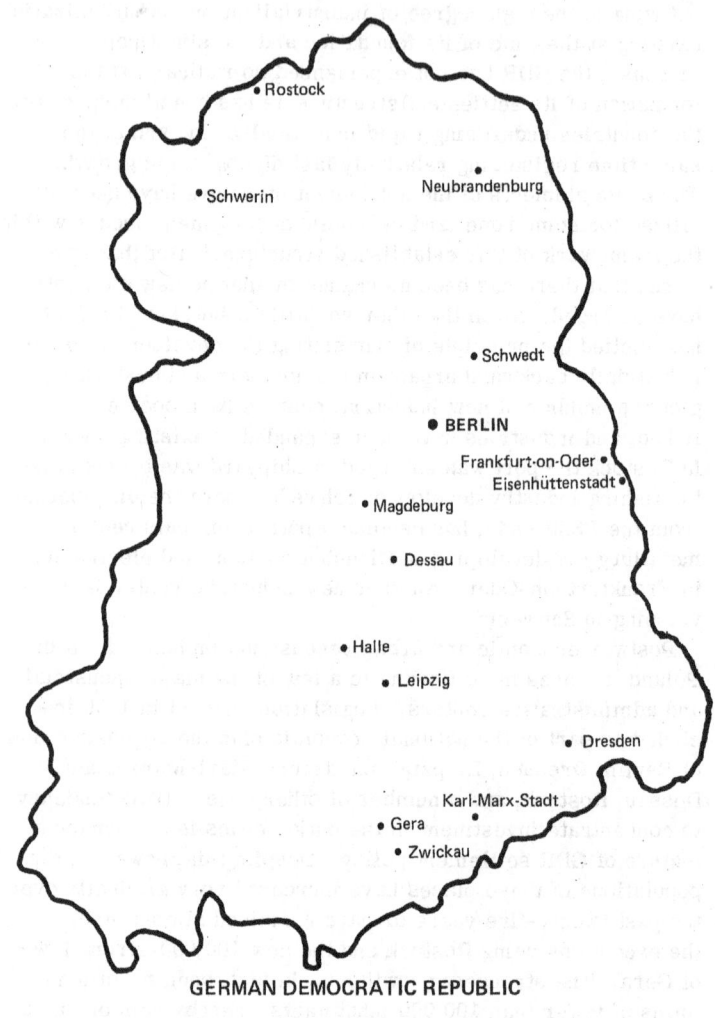

The German Democratic Republic

Table 18

Population by Sizes of Communities in the GDR, 1971

Sizes of communities	Percentage of population
-500	7.5
500-999	9.5
1,000-1,999	9.3
2,000-4,999	11.8
5,000-9,999	8.4
10,000-19,999	9.5
20,000-49,999	15.3
50,000-99,999	6.8
100,000-499,999	9.2
500,000 +	12.7

According to the 1977 figures, nearly half the GDR population (47.3%) lived in 114 towns with more than 20,000 inhabitants, and of these, one quarter lived in cities with more than 100,000. That places the GDR behind most of the other socialist countries as far as the proportion of the population living in large cities is concerned. The concentration in large cities is higher in the USSR, Poland, and Hungary.

The statistics reveal considerable similarity to the urbanization structure in Czechoslovakia, particularly with regard to the strong concentration of urban population in towns with fewer than 50,000 inhabitants. A further analogy lies in the large number of the smallest towns with populations under 10,000, of which there were 930 in the sixties, and in the large number of rural settlements. According to the 1971 figures, over 7,900 towns have fewer than 2,000 inhabitants, and 4,400 communities are below even 500. Moreover, many people living in the rural communities are not engaged in farming but travel to work in towns. In 1971 one third of the employed population traveled to work outside their place of residence.

However, the statistical view alone is not enough to reveal the nature of the settlement structure in the GDR. The small and medium-sized towns in which people live are often parts of industrial agglomerations with city cores where growth is not taking place. The German Democratic Republic is a country of industrial urban agglomerations, for over 40% of its entire population lives in five of them, their cores being Berlin, Leipzig-Halle, Karl-Marx-Stadt–Zwickau, Dresden, and Magdeburg. About 15% of the country's territory accounts for roughly 60% of all economic output. Polish agglomerations, although they occupy approximately an equal share of the territory, do not include so great a concentration of economic and human potential, and Czechoslovak conurbations are for the most part smaller.

A characteristic feature of GDR settlement is the contrast between north and south. A. von Känel and D. Scholz[3] refer in this connection to "two strongly contrasting macroareas." The northern, relatively homogeneous area is predominantly agricultural with a certain number of minor industrial regions. It also includes the Berlin agglomeration and the resorts along the Baltic coast. The south is far more differentiated; it encompasses four industrial urban agglomerations and most of the country's industrial regions, and there are also agricultural-industrial or industrial-agricultural sections. The social and cultural differences between these two macroareas are, however, minimal, thanks to the strong integration of the economy and of settlement throughout the GDR and to the spread of modern agriculture.

Settlement Strategies

Two periods can be distinguished in the development since 1949 of settlement strategies relating directly, or at least indirectly, to the principles of managed urbanization. Attention was focused during the first twenty years on optimal location of industry countrywide and within industrial agglomerations. After 1970 the approach to managing settlement broadened and was directed to setting goals and finding instruments for establishing an integrated settlement system on a national scale.

The German Democratic Republic

A sign of the increasing importance attributed to programming the urbanization process was the founding of two new institutes around 1970; they were the Institute of Geography and Geoecology, under the GDR Academy of Sciences, and the Research Station for Territorial Planning (Forschungsleitstelle für Territorialplanung), attached to the State Planning Commission. The first institute is concerned with research into the theory of settlement systems; the second is the center for applying research findings to regional planning.

The theory underlying the newly formulated concepts increasingly draws on analysis of existing settlement systems. They are divided into two basic categories: those with the function of meeting social aims and those concerned with economic aims.[4] The two types are interlocked at various levels of the national settlement system.

The long-range strategy for settlement development, which was formulated in the GDR mainly in the second half of the sixties, differs somewhat in character, both in content and in methods and form, from the Czechoslovak, Polish, and Hungarian strategies. The differences in content stem from the high degree of urbanization and, in particular, the economic importance of the big industrial agglomerations. The concept is not drafted as a national physical plan but rather in the form of general principles and guidelines.

In its overall political and socioeconomic aspects, the long-range program for settlement follows the principles of "locating productive forces" that have been formulated by congresses of the GDR's Socialist Unity Party. The program and the principles are elaborated in greater detail in special studies, which are usually the work of regional planners, economists, and geographers from the two institutes mentioned above.

General principles for industry location were first formulated at the Eighth Congress of the SUP in 1971. The resolutions included the setting of economic targets for regions. The Ninth Congress, in 1976, again dealt with industry location and, in that connection, with general principles for regional development. The basic aim, the congress decided, was to improve the territorial structure and encourage economic growth by rational location of productive forces. Specifically, it was necessary to make more efficient use of resources in all parts of the country in order to increase output and raise living standards. Progress could be achieved by utilizing existing production assets more effectively, rationalizing production in the existing industrial agglomerations, developing manufacturing industry, especially in the traditional industrial areas, and in general, carrying out so-called spatial rationalization in all districts and towns and concentrating housing construction and the social infrastructure in the first place in the cities that were natural centers of the working class. Other instruments included ensuring that the growth of cities also led to improvement of services for people living in their vicinity, progressive improvement of the infrastructures in selected small towns and villages, linked especially to advances in industrial methods in agriculture, and modernization of building stock.

All these goals have to be met in a situation of ongoing intensification throughout the economy, stagnation or possibly decline in population, further reductions in the available work force, and the need to solve the housing problem by 1990.[5]

In these circumstances specialists in the GDR regard urbanization as an essential condition for intensive economic development, and consequently they intentionally concentrate economic activities and population as a means to raise efficiency. The condition for efficiency is not just concentration in itself, however, but improvement in the infrastructures of cities and industrial agglomerations.

With a stagnating or declining population, the further growth of cities, which is one side of the urbanization process, will lead, according to the prognostic studies, to a drop in the population of small towns and rural communities. This process has already been going on for some time. Thus, according to K. Scherf,[6] between 1964 and 1975, 80% of the growth of towns with more than 20,000 inhabitants came from migration from small towns and rural communities, and 90% of their loss in population was due to outward migration. In the long run, however, conditions in the large and medium-sized towns should improve enough to make them capable of natural reproduction and not primarily dependent for their growth on the countryside and the small towns. That is also one of the strong reasons for improving the social infrastructure and environment in the cities and agglomerations.

Analysis of the literature indicates that the means for achieving the above economic and social goals through managed urbanization can be divided into three strategies. The first two are concerned with macroregional policy; the third is a central-place concept that applies to the entire territory, with certain modifications for agglomerations. First, we will examine the macroregional policies.

The planned shaping of the spatial structure in the GDR has two main objects: to utilize and develop the economic, social, and cultural potentials of the existing industrial urban agglomerations in the south and of Berlin, and second, to narrow the economic and social gap between the south and north. Underlying this policy for regional development is the need to gear the goals of a socialist society that is trying to abolish regional discrepancies while also distributing its productive forces to the best advantage to a settlement structure inherited from the past. This structure consisted of two macroareas, the industrial south, with its large agglomerations, and the agricultural north and northeast, with an inadequate technical and social infrastructure.

Current views on how to deal with this situation in the long run agree that the best way to improve the territorial structure is to use existing structures and to let the labor force and the forces of production operate where they are. It follows that in the course of improving the territorial structure in the GDR, it is not possible to develop the more backward areas at the expense of the existing agglomerations. Many authors believe it is more effective to reap the economic benefits from them and then to channel their contributions to the national income into the backward areas.[7]

Therefore the structural improvement must consist of progressively raising the level of industrialization in the northern regions while simultaneously modernizing and restructuring the existing agglomerations. Both will be long-term processes. H. Lüdemann and J. Heinzmann[8] point out that between 1955 and 1968, the share of the northern regions of Rostock, Schwerin, Neubrandenburg, and Frankfurt-on-Oder in the country's industrial output rose from 23.2 to 28.5%. W. Ostwald[9] demonstrates by other data that the process took place in the seventies. Between 1970 and 1975 the increments in gross output from the above regions amounted to 138-45%, whereas in the southern industrial agglomerations they ranged between 123 and 128%. In no way, however, should the process be seen as "levelling"; it is rather a new form of territorial division of labor.

While maintaining the prime goals of territorial development, it is necessary to make the fullest use of the advantages offered by the industrial agglomerations, which are, according to Lüdemann and Heinzmann, "...from the societal standpoint and under the socialist conditions in the GDR, the effective form for organizing the social reproduction process on our territory."[10]

The existing agglomerations provide a good foundation for intensification and complex rationalization of production, and in the future they will be the bases for further industrial development. In addition, it is clear that another agglomeration will soon join the present five, namely, Magdeburg. The advantages flowing from dense networks of settlement, industrial plant, and technical and social infrastructure must be used wherever they occur. Agglomerations can use their advantages to the benefit of society as a whole, and Ostwald lists among them: concentration of a skilled working class; links between industry, science, labor force, infrastructure, settlements, and resources that make possible better industrial performance; opportunities for contacts between industry, science, and education that will contribute to increasing productivity and efficiency in the economy; a high level of urbanization combined with concentration of educational, health, and cultural facilities serving to intensify social life; considerable regional availability of popula-

The German Democratic Republic

tion within the territories of the agglomerations, which increases their flexibility; and finally, the means to create good working and living conditions thanks to complex and varied amenities.

The large agglomerations also have some disadvantages, the most common being, according to the GDR experience: shortages of some resources, particularly labor power, building capacities, water, and sites; the concentration of old residential and other buildings; and finally, deterioration of the environment. Despite these still existing shortcomings, positive opportunities predominate in agglomeration development, and with "territorial rationalization" they can increase efficiency throughout the economy.

On the whole, neither deglomeration nor any great expansion of agglomerations is to be expected in the GDR; there is more likely to be a move toward improving their internal structures and general conditions.

The recommendation for the other parts of the GDR, where agriculture is strongly represented, and for the rural areas is to promote all methods of integrating the smaller communities and towns: for instance, making the town centers more accessible to the surrounding communities, providing technical and social infrastructures to serve several communities, and also establishing groups of communities that would pool their resources for investment, and so on. An important part in developing the settlement structure in the northern regions is played by the modernization of agriculture and the food industries, which will supplement the already existing new industrial centers. Of some considerable importance will be the strengthening of the social infrastructure, especially education and cultural facilities, which will become more accessible thanks to better public transportation services. The benefits for settlement structure deriving from the growing attraction of the Baltic coast as a recreational area are also significant. The many new facilities appearing there act as an urbanizing factor and create a whole range of job opportunities even outside the immediate area.

In the second half of the sixties, a third settlement strategy was added to the existing macroregional approach and the concepts stressing the advantages of agglomeration development. It can be described as attempting to create a "rational settlement system." Its aim is to provide conditions of life as similar as possible for everyone by means of a hierarchical system of central places. The idea is similar to the normative theories of central places. The idea is similar to the normative theories of central-place hierarchies formulated in the sixties in step was to classify settlements and delimit their hinterland zones. These zones are integrated with the centers by the movement and interaction of people, and the concept envisages the entire territory of a country divided into such city-hinterland regions. It is the function of such regions to satisfy basic needs, such as work, housing, education, retail outlets, recreation, etc. The city-hinterland regions can be interpreted as subsystems of the countrywide settlement system or as regional (local) settlement systems. The central places, i.e., the centers of regions or smaller areas, are components both of the so-called social superstructure system [11] and also the economic macrostructure. This division corresponds to the analytical distinction between two basic categories of settlement systems, i.e., those with predominantly social aims and those with predominantly economic, i.e., production, aims. The social superstructure and economic macrostructure centers are then variously classified in several levels. The best-known is that by Grimm and Hönsch,[12] which finds it expedient to distinguish six categories of centers:

— national center, capital city;
— main centers, with many specialized functions;
— provincial centers, with specialized functions for whole provinces;
— regional centers, with functions for areas larger than districts;
— district centers, with functions for whole districts;
— district subcenters, with some functions for whole districts;
— local centers, with functions for neighboring communities.

The economic macrostructure and social superstructure

The German Democratic Republic

centers should be focal points for investment and should serve both to eliminate remaining inequalities in life chances and to spread the urban way of life. The centers are the most suitable places for industrial investment and for situating scientific and research institutes. Their populations should be at least 30,000. The number of central places to be included in "superstructure" or "macrostructure" varies according to the basis and purpose of the classification. Where the accent is on satisfying needs and on easy accessibility (not more than 40-60 minutes by public transport), the number of centers selected increases; where the production function is stressed, the number declines. In practice there is always some merging and compromise between the two approaches. But an examination of the literature dealing with settlement development in the GDR shows that the number of centers scheduled for expansion has increased somewhat in recent years. This trend distinguishes the GDR theories to some extent from the concepts that emerged in Poland and Czechoslovakia during the seventies and from some of the Hungarian urbanization strategies.

The process of concentrating activities, infrastructure, and population in selected centers will be supplemented and modified on the macroscale by a tendency to establish nodal-belt settlement structures. Some writers, e.g., Mohs, Schmidt, and Scholz,[13] in common with the Polish regional school, see in the development of such a belt agglomeration structure the possibility of intensifying the urbanization process without the risk of hypertrophic growth of individual towns and deterioration of their environments.

The GDR's national settlement system is stabilized at the present time, and the forecasts for coming decades envisage no great changes on the macro- and mesolevels. On the other hand, it is probable that more pronounced changes will occur on the microscale of rural settlement. Ongoing concentration of industry in towns, and especially the industrialization of agriculture linked with the enlarging of and cooperation among farming production units, will concentrate population in fewer rural settlements. Most rural communities will have declining populations, and the more remote hamlets and small communities will gradually disappear. As a result of changes in technology, organization, and administration of farm cooperatives, small towns in rural areas will become the economic and social centers of local microregional settlement systems, i.e., centers of administration, management, and special services to agriculture, and more farm personnel than before will live in them. Their populations will rise at the expense of rural communities.

To summarize, the following trends can be expected in GDR urbanization in the coming decades: growing emphasis on concentrating economic activities, population, and infrastructure in five major industrial agglomerations and in cities; increased importance for selected settlement centers that belong to the economic macrostructure and social superstructure; movement in rural microregions toward concentration of institutions and people in fewer local centers and to depopulation of rural communities; and finally, greater integration of the whole settlement system at national, regional, and local levels.

5

HUNGARY

In 1949 half the economically active population of Hungary was still working on the land, and only 22% were employed in industry.[1] The figures are roughly the same as those for 1910 and 1930, indicating that throughout the first half of the twentieth century, Hungary remained a predominantly agrarian country with underdeveloped industry. The economic structure was matched by the settlement network, in which rural settlement predominated with, in addition to the farming communities, small and larger towns described by Hungarian writers as rural; in strong contrast there was Budapest, the metropolis of Eastern Europe after World War I. The city, which prior to 1918 had been the capital of a much larger territory with several regional centers (Bratislava, Košice, Oradea, Cluj, Tímisoara, Subotica) on its fringes, became a focal point of urbanization after that date. Between 1921 and 1941 its population increased by nearly half a million, to which should be added an increase of 260,000 in agglomerations beyond the city boundaries. Such rapid growth was bound to mean that other Hungarian towns were held back, and that the relative share of Budapest in the total population and the urban population of the country was greater than before. With its 1,712,451 inhabitants in 1941, Budapest accounted for 18.4% of the population, and the metropolitan area as a whole, with its 1,852,000 people, was inhabited by one fifth of Hungary's population.

In the very first years of building the People's Democracy

in Hungary, major structural changes took place in the national economy; they were reflected in the social structure and to some extent in the structure of settlement as well. The period of postwar reconstruction was followed by the beginnings of socialist industrialization, with emphasis on heavy industry and energy. The process was centered on the old manufacturing and mining regions in the Hungarian Central Upland and in the Budapest agglomeration. In addition there was development of towns in the southern mining districts of the Danube Basin (Dorog), and industry was built up in such old town centers in the Danube Basin as Rábo, Székesfehérvár, Káposvar, Zalaegerszeg, and Szekszárd. In the older industrial centers, such as Pécs, Sopron, Szombathely, and Pápa, and also Miskolc, socialist industry grew more slowly in this first phase of socialist industrialization. Old provincial towns, such as Debrecen, practically stagnated. The rate of industrialization is clearly shown by the figures for industrial employment. In 1949 there were 963,500 people employed in industry; by 1960 the figure was 1,682,200. In relative terms the increase was from 21.9% of the economically active population in 1949 to 31.9% in 1960.

Since in this first phase industrialization was concentrated in the Budapest agglomeration and the Hungarian Central Upland there could be no great change in the geographical location of industry and consequently of the urbanized area. E. Lettrich states that in fact the uneven geographical distribution hardly changed at all compared with capitalist times. Improvement (in this respect) would have required, among other things, major investment in transportation equipment, which in those days of "high pressure" planning was impossible. Although several new mining and industrial towns were built (Komló, Kazincbarcika, Oroszlány, Várpalota, Dunaújváros, etc.), and some farming settlements turned to industry, Hungary's settlement structure remained essentially unchanged, as is evident, for instance, from the stability of service centers, which the Hungarian geographer Pál Beluszky has studied.[2]

Pronounced changes followed only after 1965, as a result of big investments that transformed the country's economic and

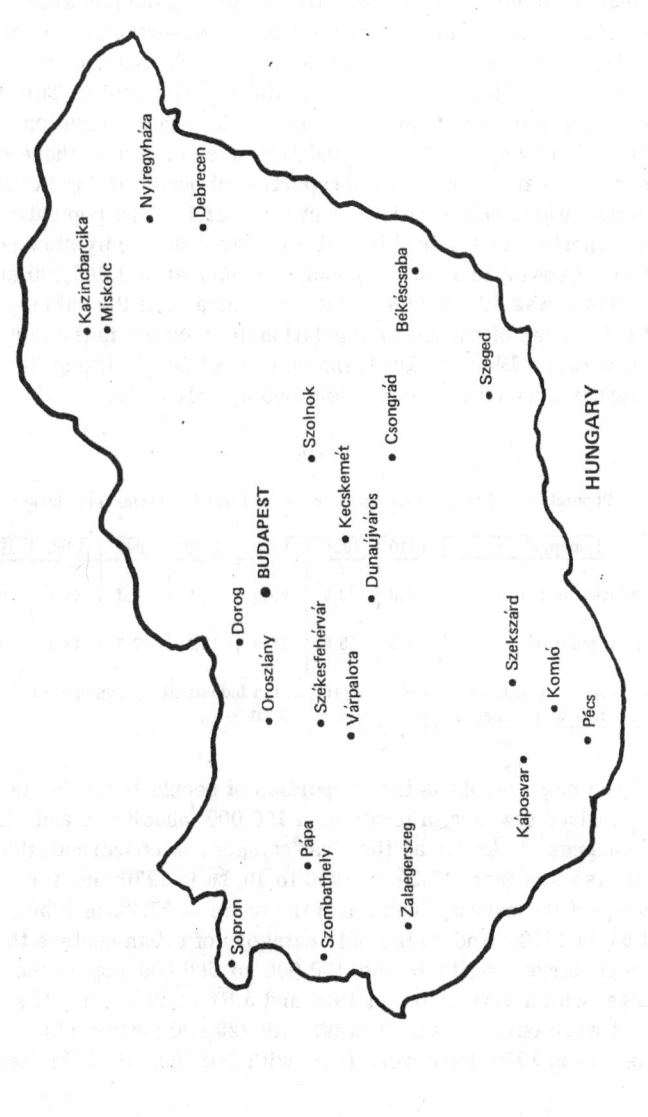

industrial structure and thereby also the settlement system. Rural settlements were radically changed by the transition to socialist forms of large-scale farming. Nevertheless, in comparison with some other countries, e.g., Bulgaria and Romania, Hungary's settlement system remained fairly stable. This is demonstrated, for instance, by figures for the distribution of population by sizes of towns and settlements. First, there has been no great change in the proportion of people living in rural communities, which still account for over half the population. An important factor in this stability was undoubtedly slow population growth; in 1949 the population numbered 9,205,000 and in 1974 it was 10,428,000. Also symptomatic of the rather slight degree of change in population distribution is the fact that between 1949 and 1974, the number of people living in typically agricultural areas declined by only 3-7%.

Table 19

Population of Budapest as Percentage of Total Hungarian Population

Budapest	1910	1920	1930	1938	1949	1960	1966
Population in industry	54.0	53.7	59.6	53.9	51.4	43.1	39.9
Total population	14.5	15.5	16.9	18.3	17.3	18.1	19.2

Source: L. Fodor and J. Illés, "Metropolitan Industrial Agglomeration," Regional Science Association Papers, vol. XXII, p. 70.

Even more stable is the proportion of people living in medium-sized towns with fewer than 100,000 inhabitants and also in Budapest.[3] As far as the first category is concerned, the increase was from 17.2% in 1950 to 18.3% in 1970; and for Budapest the relation was much the same — 17.1% in 1950, 18.8% in 1970. And so the only category of urban centers to record larger shifts is the 100,000 to 500,000 population range, with a 2.5% share in 1950 and 5.8% in 1970. In 1950 there were only two such towns, with 229,000 inhabitants, whereas in 1970 there were four, with 640,000. By 1974 there

Hungary

were five. Their rise is undoubtedly due to a decentralization policy designed to relieve Budapest, which remains, nevertheless, not only the strongest urban concentration (where, if we include Greater Budapest and its agglomerations, roughly one half of the Hungarian urban population lives) but also to an even greater extent the largest industrial center, employing in the seventies some 40% of the industrial work force. It is also a cultural and scientific center.

Table 20

Population of Budapest and Its Agglomeration, in thousands

Year	Greater Budapest	Zone of 45 agglomerated communities	Agglomeration population
1949	1,590	206	1,796
1960	1,805	260	2,065
1865	1,936	308	2,244
1967	1,968	329	2,297

Source: L. Fodor and I. Illés, op. cit.

A feature of Hungarian urbanization is the large gap between Budapest and the group of regional centers forming the second rank of urban communities. In 1970 Budapest had 1.94 million inhabitants; Miskolc, the second largest city, had only 180,581. The middle range, with 500,000 to one million inhabitants, is therefore absent. Consequently, analyses of the Hungarian settlement system repeatedly underline two weaknesses — the lack of top-ranking cities, apart from Budapest, and the lack of a middle group.

As in some other socialist countries, the growth rate for the nonagricultural population has been higher in the Hungarian People's Republic than for the urban population. That is to say, industrialization has been faster than urbanization, as can be seen, for instance, in the rapid spread of rural communities with few of their inhabitants working in agriculture but with growing numbers employed in industry and other nonagricultural sectors of the economy. Such a divergence between the industrialization and urbanization processes was bound to lead to an increase in commuting to work and to the emergence of a mixed socioeconomic category of people tied by work to the towns and by their place of residence to country life.

Despite the efforts to change the traditional pattern of industrial location and to industrialize formerly agrarian areas, and despite a successful economic policy that has made economic and social differences among regions very small,[4] the macroregions of Hungary still show considerable inequalities in their levels of industrialization and urbanization. Two thirds of the industrial population live in the Danube Valley and the Central Upland. With some simplification it can be said that agricultural communities, i.e., with over 55% of their active inhabitants tied to farming, increase from north to south and east. A continuous agricultural zone also lies today on the Great Hungarian Plain and in the southern Danube Basin. E. Lettrich, who has analyzed Hungarian urbanization, writes: "The areas that have undergone rapid urbanization lie around the capital. Connected with them are the industrial districts of the northern Danube area and northern Hungary. In addition, one can distinguish a second urbanization zone following the Tisza River and including parts of the Szolnok and Csongrád districts. The remaining territory is either weakly geared to the existing urbanization process or is entirely 'passive.' So the differences in urban character between the rapidly advancing areas and the backward ones are not diminishing; they are growing sharper."[5] This conclusion applies only up to 1960. Subsequently the engineering, light, and food industries were located to a greater extent than previously in the hitherto nonindustrialized regions. Nonetheless, one of the chief goals of regional development is still "to reduce and overcome these disproportions in the physical structure of the economy."[6]

Along with problems relating to the great concentration of economic, administrative, service, cultural, and scientific activities in Budapest and its agglomerations,[7] the Hungarian settlement system is confronted, then, with problems stemming from the variant historical development of settlement in the Danube Basin and on the Great Hungarian Plain.[8] The structure to the west of the Danube and in the Hungarian Upland belongs to the Central European settlement type; to the east of the Danube it bears the marks of the Turkish invasion in the sixteenth century. During the Turkish occupation the rural population began to concentrate in larger communities and in so-called agricultural towns inhabited predominantly by Protestants. Toward the end of the eighteenth and during the nineteenth century, this network, which was among the largest agrarian settlements in Europe, was supplemented by isolated farms lying among the fields. The area therefore consisted on the one hand of numerous villages with 5,000 to 20,000 inhabitants and on the other of scattered farmsteads. Settlement strategy was therefore faced here with exceptionally difficult problems.

Settlement structure to the west is, on the contrary, very similar to that in Croatia, Lower Austria, and Burgenland, i.e., the rural communities are numerous and relatively small. The hierarchy corresponds roughly to the Christaller settlement model for agricultural areas. The northern section, especially the Central Upland, is notable for the large number of small mining settlements and towns, which in the future will face the problem of function. The concepts for developing settlement in Hungary have had to take all these problems into account, but they have always benefited from the advantages of small territory and a relatively dense urban network.

Urbanization and Settlement Strategies

Hungary is among the socialist countries that worked out a national policy for developing their settlement networks at a fairly early date, the first coming at the end of the fifties and beginning of the sixties. The reason Hungary was able to gain such a lead over the other Eastern European countries (Poland excepted) was probably that the settlement structure issues facing

Table 21

Population by Sizes of Communities,
Hungary 1970

Sizes of communities	Percentage of population
-500	2.2
500-999	5.8
1,000-1,999	11.3
2,000-4,999	17.3
5,000-9,999	10.1
10,000-19,999	9.6
20,000-49,999	11.5
50,000-99,999	5.8
100,000-499,999	7.0
500,000 +	19.4

it were serious and needed to be tackled promptly. Of prime importance was the disproportion between Budapest, with its excessive concentration of economic, administrative, and service functions, and the other cities and towns, which accounted for only half the urban population. Moreover, the country's history advanced such issues as uneven settlement development on the Great Hungarian Plain, the stagnation and decline of small towns, and the wide gap between town and country. Postwar reconstruction and the first years of building socialist society brought no substantial change in the unsatisfactory features of the settlement network; indeed, further investment in Budapest industry resulted, despite the simultaneous development of industries in other areas, in the capital city's share of industrial output being 42.2% in 1955, with 44.1% of the industrial work force living there.

With a view to moderating and gradually eliminating these inequalities in the physical structure of the economy and in living standards in different parts of the country, the Design Institute for the Physical Planning of Towns and the Budapest institute of the same name worked out, in 1958-62, "A Study of the Development of the Settlement Network,"[9] which was based on very thorough surveys and analyses of the situation in the country as a whole and in its various areas. This resulted in a detailed proposal for future settlement. It is interesting to see what aims the study pursued, both explicitly and implicitly. Essentially it was a combination of economic and social goals, i.e., an attempt to make the whole economy more efficient on the one hand by concentrating investment and on the other by gradually raising the living standard and the accessibility of amenities in all regions.

To these ends the proposal was to divide the country into nine areas, eighty subareas, and 1,030 hinterland zones for the local rural centers. The area centers would have populations of 150,000 to 200,000; those of subareas, 30,000 to 60,000; and the rural centers, with their hinterland zones, 3,000 to 5,000. The study included a recommendation that counterweights to Budapest be formed out of five other cities, namely, Miskolc, Debrecen, Szeged, Pécs, and Rába. This was in line with the government decision of 1960 to decentralize Budapest's industry. Locations were also chosen for twenty-four towns, in addition to the above, where new industry would be located, and settlements for priority development were selected on the Great Hungarian Plain.

The industrial decentralization policy for Budapest was not purely economic but also social in purpose. It was intended to slow down, even possibly halt, the city's growth, thereby improving its living conditions and, not least, providing a basis for growth in other cities. However, experience after the government decision to stop industrial expansion in the capital showed that such administrative measures to restrict urban growth cause many problems. Above all, it appeared that in the 1960s, industry was no longer the only sector causing the urban population to grow. At that time the tertiary and even the quaternary sectors were the ones showing rapid development and finding the most favorable conditions in Budapest. Contributory factors were the rise in living standards, growing demands for skilled personnel, the spread of tourism and foreign travel, and a number of others. The result was a sharp increase in the numbers employed in all types of services, management, science, and culture. In addition, subsequent analysis of the structure of Budapest industry revealed that a considerable part was composed of sectors making products that were essential for further industrial development throughout the country.[10] This applies particularly to various products of the engineering and electrical engineering industries, the output of which had to be expanded, thereby multiplying industrial undertakings in the Budapest agglomeration. It is evident, then, that restricting industrial growth in Budapest and its agglomerations is fraught with dilemmas, and an indiscriminate approach could endanger the economic growth of the whole country. So although the concentration of industry in the capital was slowed after 1960 and population growth within the city boundaries was restricted, the agglomeration as a whole continued to grow. The statistics show that as the city's population growth slowed, population increase in the surrounding towns accelerated. In short, the Budapest phenomenon is highly complicated, and there can be no doubt that every settlement policy in Hungary will have to grapple with it.[11]

Second-generation Hungarian proposals for settlement development followed a government decision in 1971 that called for public service amenities to be equally distributed throughout the country by providing the proper equipment to enable all citizens to enjoy the same standards. Such a demand was out of step, however, with a settlement structure marked by so many local and regional disproportions, the overgrowth of Budapest, and the sparse network of other cities. The Research Institute for Town Building and Planning (VÁTI) therefore produced its "National Concept for Developing the Settlement Network to the Year 2000." Work on it was completed in 1973, probably in a first draft.

Some material changes in the draft were contained in the Hungarian document presented at the "Habitat" Conference in Vancouver, concerning the number of settlements in the different categories. But the basic principles have not altered. The present writer considers that somewhat greater weight was given to concentration of production and thereby, in part, of population. The most recent information is that work on the concept is continuing, and Hungarian regional planners are trying to introduce some interesting innovations into it. For instance, they are considering whether a strict hierarchic grading of functions within the settlement system could be avoided. Specifically, they are examining the possibility of not always assigning the lower-, middle-, and higher-level centers all the so-called central functions that should belong to them, but perhaps locating some of these functions in parallel centers that would then form with the main center a quasi group system of settlement. Essentially this is an attempt to suppress the rigid hierarchic principle that is often applied in the case of the central-place theory in particular. But let us now return to the first version of the concept dating from 1973.

Its aims have been summarized by K. Perczel, one of its authors: "...It was necessary to propose a system of urban centers and their hinterland zones in which every citizen, wherever he lived or worked, would share equally in modern public facilities at all levels, these being at readily accessible distances. In such a system of urban centers and their hinterlands, the lower-level centers must, of course, be established in the immediate vicinity of each settlement. The higher-level centers can be developed at geographically more thinly distributed points linked to a suitable transportation network at the most easily and quickly accessible places or at transportation junctions.... Given the present urban network in Hungary, such a system of centers has not been evolved with a socialist character capable of providing full service to citizens, the main reason being the inadequacy and weakness of two types of center. One weakness relates to the higher-level centers of special significance...the second is the small number of middle-level centers and the consequent inadequate service to areas where towns are lacking."[12]

We have quoted this long passage not only to indicate the lines the concept follows but also to show the important role attributed by the Hungarian experts to public services. That does not mean, however, that the VÁTI model is primarily a "consumer" model. One integral part of it is also a thoroughly worked out section dealing with suitable locations for new industrial plant and for industrial development in general.

All in all it can be said that the Hungarian concept aims to strike a balance between a settlement model based on promoting industry and one motivated by the principle of service to the population. The document prepared by government institutions for the UN conference on settlement, held in 1976 in Vancouver, principally emphasizes the connection between settlement strategy and the development of productive forces, but otherwise it mainly stresses objectives belonging to the "consumer" sphere. During the seventies in Hungary, as in the other socialist countries, there has been emphasis on the need to proceed from extensive to intensive forms of economic development.[13] This change lends added importance to the territorial distribution of industry. The document for the Vancouver conference states: "At the present time, in connection with the transition to intensive forms of development, the territorial aspect of industrial advance is gaining in importance. The government stresses resource utilization, rational territorial concentration and bringing labor-intensive industries close to the sources of labor...."[14]

One instrument for eliminating the disproportions in the settlement structure, improving public services, and achieving rational localization of industry is a categorization of centers in four basic levels, divided as follows:
— national centers;
— higher-level centers;
— middle-level centers;
— lower-level centers.

The capital city, with the status of a national center, performs three main functions. It is the service center for the 2.5 million inhabitants of the Budapest agglomeration; it has statewide functions for all the Hungarian population and also has an important international function.

The cumulation of central functions in a single place requires as a counterweight the establishing of regional centers

Hungary

"of a level, complexity, and degree of specialization enabling them to approximate the level of the capital." Miskolc, Debrecen, Szeged, Pécs, and Rába are intended to meet this need. The concept for these development poles rests on a careful selection of modern branches of industry suited to them. In addition to these five special-significance centers, the concept envisages another seven centers in the higher-level category (Székesfehérvár, Szombathely, Szolnok, Káposvar, Kecskemét, Békéscsaba, and Nyíregyháza). At a later stage in the work the number of higher-level centers was increased from seven to eighteen. Middle-level centers should number something over a hundred in a thirty-year view, those of lower level around 900. A later version gave a somewhat lower figure. We should add that in addition to the invariant forecasts for settlement development, the Hungarian Planning Commission has produced three variants for possible regional policies. They differ in the degree of concentration of economic activities and population and the greater or lesser dispersal of towns.

The 1973 national strategy also envisaged an urbanization rate measured by the concentration of population living in urban communities. By 1985 the urban population was to rise to 54% of total population, and by 2000 to 64%, with the suburban zones advancing to 72-75%.

Comparing the "National Concept for Developing the Settlement Network to the Year 2000" with "A Study of the Development of the Settlement Network," published ten years earlier, we find that the Concept is really a modification of the Study.[15] Although the new version was criticized in discussions both for assuming too much concentration and for leading to excessive decentralization, as well as for having too many middle-level centers, there is no question that one can observe some shift toward stressing concentration of population and some change in the view of Budapest's development. The number of higher-level centers has been increased in the new proposal and that of the lower-level ones reduced. The "principle of relieving" Budapest is still applied, but decentralization policy is seen as selective and is not meant to touch the most important sectors of Budapest industry, for that could retard economic growth in the whole country. New industries are to be concentrated as far as possible in selected industrial centers; and, as G. Enyedi stresses, modern agriculture should be vertically integrated, i.e., forming larger, complex units can fulfill an urbanization role in rich farming areas by situating supralocal facilities (administration, services, storage, specialized agronomy centers) in a modern version of the agricultural towns. This concept certainly gives more weight than the earlier study to forming agglomerations and to "decentralized concentration" as the central idea in the country's settlement policy.

6

ROMANIA AND BULGARIA

Romania

Before World War II Romania was among the backward agrarian countries on the European continent, with weak industry and raw material extraction developed in but a single direction. There was a settlement structure to match, marked by a low concentration of population in towns and the complete predominance of rural communities. In 1930 only 21.4% of the population lived in urban communities, and by 1948 little had changed, the urban element having risen to a mere 23.4%. The turning point came, as in many East European countries, with the onset of socialist industrialization. The first phase, in particular, was notable in Romania for rapid increase in urban population. The annual urban growth rate between 1948 and 1956 was reminiscent of the USSR in the thirties, with the annual increments being on the order of 5%. The later phases of urbanization saw a moderation in the rate, but it remained fairly high, ranging until 1970 from about 2.8 to 3.1%.[1]

Romania is, then, a country that has undergone profound changes in its economic and settlement structure, changes associated with extensive migration from the countryside to the towns and the growth of commuting to work in urban industrial centers. Over the short period from 1948 to 1969 the urban population increased by 4,382,000, which was practically the total population growth. Although in absolute terms the number of people living in the countryside remained more or less the same over the period, i.e., at 12 million, the change was evidenced by a sharp drop in the share of the rural element in total population; in 1948 it was 76.6%; by 1969 it had sunk to 59.9%, and the rapid decline can be expected to continue. Moreover, many of those who have stayed in the country have transferred to industrial employment and travel to nearby factories. In the decade 1956-66 alone the number of industrial workers living in the country rose from 1.5 million to 3.2 million, and this special urbanization process is undoubtedly continuing now.

Table 22

Rural and Urban Population of Romania, 1930-69

Year	Total population, in thousands	Rural population in thousands	in %	Urban population in thousands	in %
1930	14,280.7	11,229.5	78.6	3,051.2	21.4
1948	15,872.6	12,159.5	76.6	3,713.1	23.4
1956	17,489.5	12,015.2	68.7	5,474.3	31.3
1966	19,103.2	11,797.4	61.8	7,305.8	38.2
1967	19,248.8	11,816.7	61.3	7,468.1	38.7
1968	19,720.9	11,812.1	59.9	7,908.8	40.1
1969	20,010.2	11,914.0	59.5	8,096.2	40.5

Source: C. Rusenescu, "L'urbanisation et les nouveaux rapports entre la population, le site et le territoire de villages roumains," in IGU, Urbanization in Europe, Budapest, 1975, p. 91.

The migration flowed to towns of varying sizes, but a characteristic feature of Romanian urbanization in the past twenty-five years has been a rapid increase in the proportion of the population living in towns with more than 100,000 inhabitants.[2] In large measure the process can be explained by the fact that a number of small towns had passed the 100,000 mark (in 1950 there were four; by 1970 there were thirteen), but it was also a result of the Romanian government concentrating industrial investment in these towns. Therefore the traditional industrial centers, such as Bucharest, Brașov, Ploiești, and Hunedoara, grew, and there was rapid growth of towns like Craiova, with its newly established industry. Naturally investment went not only to the 100,000-strong cities but also to smaller places, for instance, to Botoșani and Piatra Nemt. This regional investment policy has put Romania in first place among the socialist countries in developing the so-called metropolitan forms of urbanization, i.e., concentrating population in towns with more than 100,000 inhabitants. It also has the highest figure for city density; but as V. Matoušek[3] points out, the majority of Romanian towns, apart from Bucharest, lack technical infrastructures comparable to the standards found in towns of the same size in Czechoslovakia or the GDR, for instance; moreover, they are not cores of industrial urban agglomerations, as is the case, for example, in the GDR and other socialist countries. It is, then, predominantly a so-called point form of urbanization, although the rudiments of agglomerations certainly exist, for instance, the twin towns Brăila-Galati, which with their surroundings approach a half-million urban population concentration. Analyses by P. Deica and J. Stefănescu[4] indicate the existence of agglomerations, but it seems — and data on population density in these settlement formations suggest — that in many cases they are more in the nature of emergent agglomerations. In 1973, 7.6 million people were living in twenty-one agglomerations identified by Romanian writers, 35% concentrated in the two largest centers, i.e., the Bucharest agglomeration (1.7 million population) and Ploiești (900,000). A further six agglomerations had 300,000 to 400,000 inhabitants; in eleven there were 200,000 to 300,000 people; and in the two smallest, fewer than 200,000. The majority of Romanian agglomerations have low population densities, i.e., not reaching 150 people per square kilometer. Another indication of their being emergent formations is the fact that so far the most rapid growth has been in their inner areas, the urban cores.

Bucharest's development is interesting because during the first years of socialist construction, its growth and its relative

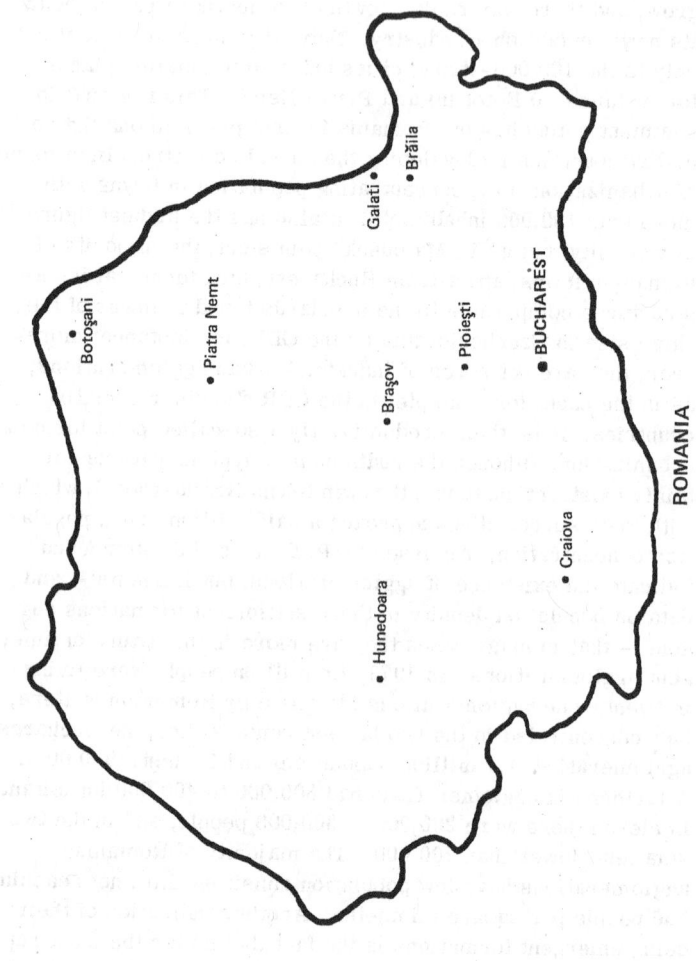

importance rose rapidly. According to the 1956 figures, one quarter of the country's entire urban population lived in the capital; later, undoubtedly due to the policy of developing other cities, its share began to decline. Although the population of the Bucharest metropolitan area increased between 1950 and 1970 from 1.1 million to 1.7 million, its share in the urban population dropped to 19.9%.

Table 23

Proportion of Population in Towns by Size Groups and Its Growth in Romania, 1948-70

Community sizes	1948	1966	1970	Indexes		
				1948 to 1966	1966 to 1970	1948 to 1970
up to 10,000	11.3	6.6	7.8	98.4	141.0	138.8
10,000-25,000	20.7	19.9	21.1	162.6	126.5	205.7
25,000-50,000	10.6	14.3	13.9	228.9	115.1	263.5
50,000-100,000	22.9	8.6	10.6	63.6	147.1	93.6
over 100,000	6.2	28.7	26.7	777.1	110.8	860.8
Bucharest	28.3	21.9	19.9	131.1	107.4	140.9
Total	100.0	100.0	100.0	169.1	118.9	201.2
Total urban population in thousands	3,678.3	6,222.9	7,402.9	—	—	—

Source: C. Rusenescu, op. cit., pp. 92-93.

The problems connected with developing settlement in Romania are many and quite considerable, as is inevitable with such a rapid transition from an agrarian-feudal socioeconomic structure to an industrial socialist state, and with the type of settlement structure that was inherited — a structure belonging for the most part to the sparse Eastern European type which, in addition, had its development disrupted by numerous discontinuity factors, for instance, the long period of Turkish rule. Fragmentation of settlement is typical. Following the administrative reform of 1965, the country has around 13,000 villages and something over 200 towns. Over two thirds of the villages have fewer than 1,000 inhabitants. Consequently the foremost problem for the future is to merge communities and create larger settlement formations.

With the exception of the Carpathian mountain and foothill areas and the Danube Delta, population density is fairly even. Another good feature of the settlement network is that the regional centers are, on the whole, symmetrically distributed. A problem, however, is the scarcity of medium-sized towns to form some kind of link between the cities and rural settlements. In view of the country's strongly agrarian character, macroregional differences are largely caused by the types of farming products. Along with rich agricultural areas from which fewer and fewer people are moving to the towns, there are the backward and poor pastural areas of the Carpathians, for instance, the Vrancea district. They are being rapidly depopulated, and therefore a more suitable development strategy has been sought for them. There are also some differences in settlement structure and in some sociocultural features between Transylvania and the rest of the country.

Despite the rapid growth of towns and the formation of metropolitan areas, the most serious problem in developing the Romanian settlement network remains the organization of rural settlement and linking it to existing urban centers.

Bulgaria

For some time now Bulgaria has been among the European countries where urbanization has been proceeding most rapidly. The rate is a direct result of socialist industrialization carried out in the form of building large works and plant complexes and of socialization in agriculture, which has released labor for other sectors of the national economy. This country, which in the prewar era was completely agrarian, started to develop its industry with only one quarter of its population living in urban communities. In 1950 rural population still amounted to 72.6%

of the total, but in the course of twenty years this proportion has declined to the extent that today over half the Bulgarian population lives in urban settlements. The rate of this decline in rural population is comparable with that recorded in the Soviet Union and, to some extent, in Poland.[5]

This economic and settlement transformation stemmed, among among other things, from an increase of 2 million in the urban population between 1950 and 1970, from 1,987,000 to 3,982,000, and an accompanying decrease in the countryside from 5.3 million to 4.5 million. This meant that the towns absorbed not only the entire population growth of the country but also the 800,000 reduction in the rural population. In other words, urbanization in Bulgaria was more rapid than in Romania, Yugoslavia, Hungary, Czechoslovakia, and the GDR, and there was not the extensive growth of a category of nonagricultural rural population found in those countries.

The migration from the countryside flowed mainly to cities that are today in the 100,000 to 500,000 group and to Sofia. Between 1947 and 1970 migration between rural communities and towns yielded a balance of 1.5 million, the majority going to large cities. The rate of urbanization is also documented by the fact that in 1950 there were only two cities with over 100,000 inhabitants, namely Sofia and Plovdiv, whereas by 1970 there were six. Their growth was particularly rapid in the decade 1960-70, the average annual increment being between 3 and 4.5%. Bulgarian studies of settlement development state that the reason for concentration in large towns is "... the growth of industrial plants that expanded rapidly in these towns owing to concentration of production and the opportunities for them to specialize and concentrate. Medium-sized towns also grow in association with industrial development.... Some of the medium-sized towns expanded because they were designated as administrative centers of districts...."[6]

As a result of this urbanization process, by the midsixties over 40% of the total urban population was living in the cities. Sofia's growth has also been remarkable. In the longer view, too, its growth rate is unprecedented in comparison with other

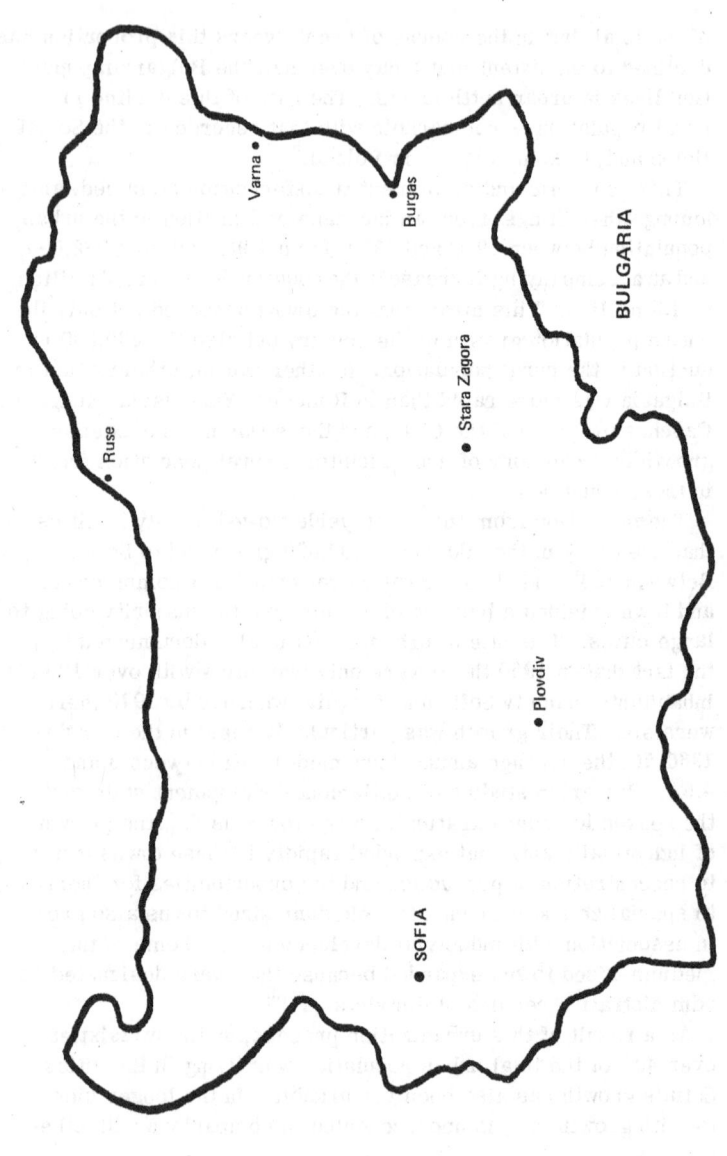

Romania and Bulgaria

Table 24

Population by Sizes of Communities
in Bulgaria, 1965

Community sizes	Percentage of population
-500	6.7
500-999	11.5
1,000-1,999	18.1
2,000-4,999	17.1
5,000-9,999	7.9
10,000-19,999	4.8
20,000-49,999	8.5
50,000-99,999	8.0
100,000-499,999	7.7
500,000 +	9.7

European capitals. In 1881 the city had a population of little over 20,000; in 1934 the figure was 287,000; after World War II there were 500,000 people living in the Sofia metropolitan area, and estimates put the present population at 1.2 million. Nevertheless — and this is an interesting feature of Bulgarian urbanization — Sofia's position in relation to the urban population is not "hypertrophic." It is estimated that at present roughly a quarter of the country's total population lives in the metropolitan area; that is a higher proportion than in Warsaw, but not much more than in Bucharest and, of course, less than Budapest. A certain balance in the urbanization process has been caused by the simultaneous rapid growth in recent decades of other cities — Plovdiv, Varna, Ruse, Burgas, Stara Zagora — and also of medium-sized towns.

Concentration of economic, social, and cultural activities has already led to urban agglomerations — already established or in course of formation — such as the Sofia-Pernik agglomeration and those of Varna, Plovdiv, and Burgas. Although their populations do not reach the dimensions of the Soviet or Polish agglomerations, by the nature of their internal relationships they are representative of newly developing settlement systems.

Urbanization in Socialist Countries

According to studies of the subject, the problems facing Bulgarian settlement for the future stem from low density, which is evident from low population density and the sparsity and fragmentation of rural settlement. There are on average 5.1 settlements to 100 square kilometers. Bulgarian experts believe the solution to these problems, which derive in large measure from factors reaching back into the country's history, lies in progressive integration of rural settlements into new regional systems.

Urbanization and Settlement Strategies

The principles underlying Bulgarian and Romanian concepts for developing the settlement system can be summarized as follows:

1. The old settlement network, which was shaped primarily by the predominantly feudal-agrarian economic structure and to a lesser extent by the beginnings of capitalist industry, rapidly comes into conflict with the economic and social structures emerging in socialist society. The settlement structure therefore has to be systematically transformed to correspond to the new social conditions as closely as possible and to prevent it being a brake on economic and social evolution.

2. The main feature to be eliminated from the old structure is dispersion and fragmentation, particularly of rural settlement. G. Sebestyén emphasizes that "the number of small villages existing in the present Romanian network will undoubtedly be a key problem for future development."[7] Bulgarian writers have expressed the same view. The course to be followed must therefore be to gradually merge villages, reduce their number, and group them in larger communities.

3. A further principle follows from the above ideas, i.e., to establish settlement systems made up of cooperating communities.

4. To restrict overgrowth of cities, especially capitals, and to stimulate growth in medium- and small-sized towns, which would become, in Gusti's terminology, "catalysts"

or nuclei of the settlement systems.

5. To develop urbanization rapidly on the basis of the medium- and small-sized towns and their agglomerations, with a view to achieving a high degree of urbanization as soon as possible.

The formation of settlement systems based on cooperating communities is a vital principle. Bulgarian and Romanian economists, sociologists, and urbanists expect a whole number of urgent problems to be solved by their planned development. They are primarily a way to help overcome the fragmentation of rural settlement and eliminate differences between town and country by "drawing" a fair number of rural communities into the urban settlement systems; finally, they are a way to restrict growth in oversize cities while maintaining a rapid urbanization rate, which is an objective process stemming from industrialization and other macrosocial changes.

Regarding the role of settlement systems of cooperating communities in relation to the rural situation, the most precise formulation comes from G. Gusti: "In a situation where settlement is based on a dense network of small- and medium-sized communities, located not far from each other, facilities and services can best be provided by cooperation among a certain number of neighboring points.... Such groupings can offer the population some of the special advantages of cities, especially the greater variety of job opportunities provided by more differentiated production and economic structures, thereby also making better and more rational use of labor power and providing the public with more comprehensive services. Settlement based on grouping together places that lie close to each other but are not joined can preserve for the inhabitants all the advantages of medium-sized towns — the human dimension, closeness to nature, healthy and comfortable living...."[8]

Group settlement systems are regarded as new and higher forms, replacing the pattern of separate and relatively independent communities. In a speech in 1972 dealing with urbanization matters, Todor Zhivkov stated that "...a settlement system of this kind represents a qualitatively higher formation both in comparison with separate towns and with groups of towns and other places that are linked administratively or industrially."[9]

Romanian efforts distinguish between first-level systems, containing the basic structural elements of today's villages, with the minimum size for the catalyst nuclei being put at 3,000 inhabitants, and second-level systems formed by urban settlements of varying sizes, which could satisfy the less basic needs of the population. In the long run, cooperation should develop among several second-level systems; thus the entire settlement network would gradually be completely integrated. The idea is very similar to that developed in Soviet settlement strategies.

It is interesting that Romanian studies, when considering the purpose of the group settlement system, tend to see that its aim of getting sufficient concentration of population is a condition for providing more advanced forms of public facilities, whereas Bulgarian works tend rather to derive group settlement systems from the economic-industrial situation and to view them as products of rational location of production. This is to be seen in L. Tonev's attitude: "When planning the comprehensive location of productive forces and population for whole economic areas and subareas ... it is possible and necessary to formulate and solve the town-planning problems not in isolation for individual towns and communities, but simultaneously and comprehensively for larger or smaller groups that constitute integral settlement systems."[10]

Similarly, in the work of the Research Institute for Town Planning and Architecture, it is stated that "... on the basis of more intensive production relations among individual towns and settlements, new settlement forms of higher quality develop." Included among them are "settlement systems of the settlement-agglomeration type," which arise on the basis of large industrial centers; "settlement systems that represent not only certain groupings of towns and other communities but also a variety of more lasting ties that determine the course and specific forms of their economic development"; and finally, "twin towns," in which both towns grow, extend their mutual ties, but still retain their independence.

Preventing the excessive growth of cities, which is the official policy of the Bulgarian Communist Party, is also to be achieved by means of the group settlement system. Todor Zhivkov has stated explicitly that "...this is one of the ways to effectively restrict city growth." It can be done by regulating the size of settlements within the systems, by building more centers, and so on. At the same time, the advantages of urban life are made accessible to people living in small communities by providing transportation links with the system's center, i.e., with one or several large towns.

Parallels between the Romanian and Bulgarian ideas are also evident in their recommendations "to build up a broad network of medium-sized towns, i.e., with populations of 25,000 to 100,000" (Grekov), and "to create a range of settlements of varying sizes in which we would regard those with 20,000 to 50,000 inhabitants as medium-sized" (Sebestyén). Moreover, in both countries it is regarded as expedient to regulate the growth of the capital cities and to try to form a balanced polycentric settlement network and avoid capital monocentrism. The Bulgarian town planners in particular stress a polycentric structure, seeing it as a way to "tear up" the traditional urban structure and advance toward belt formations of settlement. They seem to place greater emphasis on larger urbanized areas, on "development axes,"[11] than is the case with the Romanians.

Work has also been published in both countries estimating the degree to which population should be concentrated in towns by the end of the century. According to some Romanian forecasts,[12] 68 to 71% of the population should be living in urban communities by 1980, and by the year 2000 the proportion of urban population should rise to 75%. It should be mentioned, however, that the projection worked out by the United Nations envisages the urbanization process being slower than assumed by Cănilă.

In Bulgaria, where urbanization is presently proceeding at the highest rate among the socialist countries, the urban population should, according to the published forecasts,[13] rise to 8 million by the year 2000, amounting to 80% of total population. The dimensions of the social change taking place in that country in the second half of our century are obvious from the fact that when World War II ended, only a quarter of the population lived in urban communities. There can be few countries in the world where urbanization has proceeded so rapidly. That also bears witness to the dynamism of the social and economic changes Bulgaria has experienced.

7

YUGOSLAVIA

Although the urban population has grown in Yugoslavia since World War II at twice the rate recorded in 1921-31, the country remains among the least urbanized in Europe. To a large extent this is still the result of the low level inherited from the past. According to I. Ginić,[1] only 17.5% of the population was urban in 1931, and by 1948 the proportion had risen to 19.8%. An increase of over 5 million in the towns over the period from 1950 to 1970 was not enough to change Yugoslavia's position in this respect, because in the other agrarian countries of Europe urbanization had either been more rapid or at least equal to that in the Yugoslav Federation. Consequently, with 7.6 million urban inhabitants, representing 36.7% of its total population, Yugoslavia today is still in the category of the least urbanized areas in Europe. Despite the shift from 28% in 1961 to the present level, the country's position has not been altered. The view is therefore often heard that "... the urbanization level reached in Yugoslavia is not realistic in relation to the level of economic development the country has achieved. Urbanization has not kept in step with the other development processes, and the proportion of urban population is substantially lower than that generally regarded as normal for the given economic structure and the other indicators of progress."[2]

As in Romania and Bulgaria, urbanization in the Socialist Federative Republic of Yugoslavia is inseparably linked with what Yugoslav authors term deagrarianization, i.e., the transfer of part of the population from agriculture to other economic activities, particularly to industry. Characteristic for Yugoslavia is that the deagrarianization rate is higher than the rate of concentration of people in towns. An analysis of these two parallel and interdependent processes for 1953-61 shows that 60,000 former farm workers moved each year to towns, whereas 180,000 who had formerly worked in agriculture left this sector of the economy but continued to live in the country. That led, naturally, to the emergence of a large socioeconomic category consisting of a rural nonfarming population. Kosta Mihailović estimates in his Regional Development: Experiences and Prospects in Eastern Europe [3] that 30-40% of Yugoslav workers belong to the worker-farmer category living in the country. Thus they are a very important social stratum in general and in the countryside in particular. At the same time, rural population declined in the twenty years 1950-70 from 13.5 million to 12.6 million, but the relative proportion dropped from 83% to 63%.

The rapid growth of the nonfarming rural population is an objective phenomenon encountered in all the rapidly industrializ-

Table 25

Population by Sizes of Communities
in Yugoslavia, 1948-71

Community sizes	1948	1953	1961	1971
up to 300	13.0	12.3	10.8	9.6
300-600	16.4	16.0	14.7	12.6
600-1,200	22.5	22.1	20.9	17.6
1,200-5,000	24.9	24.7	24.6	22.9
5,000-15,000	7.3	7.5	8.3	9.4
15,000-50,000	6.7	7.5	8.9	11.1
50,000 and above	9.2	9.9	11.8	16.8
Total	100.0	100.0	100.0	100.0
Yugoslav population in thousands	15,842	16,991	18,549	20,523

Source: M. Rančić, "Značenije i problemy demografičeskich prognozov naselenych punktov v Jugoslavii," Liblice, 1976.

ing socialist countries. In Yugoslavia, however, it was strengthened by a number of specific conditions, among which Yugoslav authors assign primacy to the self-management system, which provides conditions for auxiliary industrial and agricultural undertakings to be located in many places, thus preventing the superconcentration of population in a few cities. However, views differ as to the benefits or otherwise of this development. A further reason for the slow population buildup in the cities is the relatively small scale of housing construction, especially during the fifties.

In the process of socialist reconstruction and industrialization, urban centers of all sizes grew in Yugoslavia;[4] but it is evident that the medium-sized towns of 15,000 to 50,000 inhabitants recorded somewhat above-average growth, thereby reinforcing the traditional element of medium urbanization that already existed in prewar Yugoslavia. The great importance of the 20,000 to 50,000 population category is evident from the fact that there were eighty-two such towns in 1970. As in most countries where small towns with fewer than 15,000 inhabitants acted in the past as service centers for the farming population around them, these places now face many problems. Having lost their traditional functions, they have not yet acquired the new ones that greater industrialization brings.

The federative constitutional order of the postwar period has also made the capitals of the union republics expand quickly; this applied particularly to Skopje, Zagreb, and Belgrade, while Sarajevo has grown rather more slowly, and Ljubljana slowest of all. Of the other cities at the 100,000 population level, Niš and Novi Sad have grown very rapidly, as have Banja Luka and Rijeka recently. The federation has undoubtedly contributed to Belgrade's primacy not being as pronounced as that of the capitals of the other socialist countries. The metropolitan areas of Belgrade and Zagreb have populations of the same order of magnitude.

Although the factors underlying the urbanization process are the same for all parts of the country, there are considerable differences between the union republics stemming from the

Yugoslavia

quite substantial variations in economic, social, and also settlement conditions and from their different historical backgrounds. Yugoslav settlement belongs, in fact, to two areas. In the northwest and part of Serbia it is of the central European type, whereas in Macedonia, southern Serbia, Kosovo-Metohija, and Montenegro sparse eastern European settlement prevails, although around Skopje and to the south of Macedonia the population density is fairly high. Settlement in Vojvodina (Banat) is similar to that in the southern parts of Alföld. These differences — stemming from historical circumstances among which should be included the fact that Metohija, for instance, and other parts were under Turkish rule until 1913, whereas Slovenia, Croatia, and Vojvodina were until 1918 part of the Austro-Hungarian Empire — have been intensified by the different rates at which industrialization and social change have proceeded in the union republics in the period of socialism. In Croatia and Serbia urbanization has come closest to the classical type, marked by the formation of large urban concentrations. In the seventies Belgrade and its metropolitan area contained about half the urban population of Serbia. Over one third of Croatia's urban population was concentrated in Zagreb. In Macedonia urbanization is running ahead of economic development, so that Yugoslav authors refer to that republic as overurbanized. A strong factor here has been the rapid growth of Skopje following the earthquake in 1963. Leaving aside Vojvodina, Macedonia has the highest urbanization level in the whole Federative Republic. The areas of lowest urbanization are Bosnia and Hercegovina, together with Kosovo, Metohija, and Montenegro. Slovenia is a special case. Although it is the most economically advanced part of Yugoslavia, urbanization, including growth in the two largest cities, Ljubljana and Maribor, has been slow in recent decades. It has assumed an indirect form — the urbanization of rural settlements. People who gave up farming did not leave the villages. Consequently, in 1961, for instance, 68% of the nonagricultural population lived in the country. The high proportion of urban population in Vojvodina, which leads the country in urbanization and is also Yugoslavia's granary, is explained by the nature of settlement there. The many farming towns that were established in the past for purposes of defense today serve as centers of a rich agricultural hinterland.

Settlement Strategies

A statewide concept for settlement development has not yet been attempted in Yugoslavia. This is certainly connected with the federative system, under which regional and physical planning are the responsibility of the union republics, not the federal government. Physical planning stems from commune needs and is hampered by a considerable lack of unity in legal norms. The development plans for Serbia and Slovenia[5] come closest to the proposals for developing settlement networks in other socialist countries. The Parliament of the Serbian Socialist Republic decided in 1967 to work out a plan for developing Serbian territory. The plan was conceived as a way to achieve general social goals by means of physical planning decisions and to coordinate the various activities taking place in the area. The work included extensive studies designed to assess the natural, economic, social, and infrastructure conditions in the republic. Basic development strategies for different parts of Serbia are laid down in the plan. Regarding the structure of settlement, the principle of a hierarchy of centers is applied. This draws on earlier work that was devoted primarily to rural settlement and that recommended the formation of so-called basic agricultural communities, with elementary public facilities, and "circuit centers," with 5,000 to 10,000 inhabitants. These circuit centers would contain, in addition to public services, some small industrial works. The idea is similar to Soviet proposals for industrial-agricultural complexes with the settlement form of mixed agricultural towns. The authors of the Yugoslav settlement concepts can draw on experience, because in 1959-64 ninety-four such complexes were established in order to eliminate differences between town and country.

Yugoslavia

As in Serbia, in Slovenia as well work was started in 1968 on a territorial development plan for the republic. Since supra-communal physical plans are not institutionalized in Yugoslavia, the concept was based on linking the plans for individual communities and on working out plans for groups of communities, especially for urban regions. In view of the special features of Slovenian settlement,[6] i.e., slight concentration of population in towns combined with a high degree of economic development, and in view of the republic's small area and relatively dense settlement, which bring the systems in touch with each other, the proposal does not envisage any great concentration in towns. It rests on the principles of central places with fairly strong representation of small towns in the projected network. Vladimir Kokole also describes it as the formation of "a polycentric concentration system"[7] and one of the forms of decentralized concentration. The concept has two main aims, on the one hand to end the excessive dispersal of population in a multiplicity of settlements, on the other to prevent concentration in a few large towns. The polycentric system should, according to its authors, contribute to people's creative participation in the self-management system and hence to developing the human personality, as well as to social equality, economic efficiency, the quality of the environment, and overcoming the differences between town and country.[8]

8

CONCLUSIONS ON SETTLEMENT STRATEGIES

The urbanization and settlement strategies that were formulated in the socialist countries in the sixties and seventies had, on the one hand, some permanent features in common that were not substantially changed during the two decades; on the other hand, they possessed specific features stemming from inherited settlement structures, differences in demographic situations, and also from the phase of industrialization and urbanization at which a given country found itself.

Permanent, Common Features

The reasons why the settlement strategies in the countries we have examined are similar in many respects are obvious. These are countries with similar socioeconomic systems based on social ownership of the means of production, and in the majority of them the basic economic mechanism is that of central economic planning for the entire country, although in Yugoslavia central planning plays a smaller part.

Some differences in regional and physical planning procedures are due to differences in ownership and the economic and organizational forms obtaining in agriculture. Others stem from the state system in different countries, e.g., the distinction between the federative systems in Yugoslavia and Czechoslovakia and the nonfederative state in Poland.

Urbanization in Socialist Countries

In all the countries, with the exception of Yugoslavia, settlement development planning is based on national economic development plans and plans for regional development of industry and agriculture. Urbanization strategies, also described as settlement strategies, gradually become integral parts of the instruments used in directing the evolution of the entire society.

The common features in the strategies could be summarized in the following way:

1. Planned and balanced development of all the components of the settlement system to avoid economic and social disproportions. This means that emphasis is placed on comprehensive planning of industry and other economic activities, transportation, housing, public utilities, and the technical infrastructure, as well as on their spatial interrelationships.

2. An endeavor to ensure rational distribution and efficient utilization of productive forces throughout the country, to make optimal use of the natural and socioeconomic conditions in different areas, and simultaneously to improve living conditions there.

These general principles are supplemented by specific objectives relating to the development of the settlement system as such. Comparison of the published theories reveals that they are, in particular:

1. A search for balance between the economic and social goals for developing society in the territorial aspect, i.e., achieving economic effects not at the price of creating social problems but, on the contrary, by attaining social goals without retarding economic growth. To some extent the search for this balance reflects the search for a suitable relation between production and nonproduction investment; it is expressed in this context frequently and in concrete terms by fixing a ratio for the degrees of concentration or deconcentration of settlement.

2. Reducing the social inequalities between the regions of a country, by direct aid to backward areas and by redistribution policies. Closely connected with this are policies for making equally accessible the public facilities that together provide people their opportunities in life (education, health care, culture, etc.).

Table 26

Projection of Urban Population Trends in the Socialist Countries,
UN Estimate to 1985

Population in thousands

Country	1970			1975			1980			1985		
	total	urban abs.	in %	total	urban abs.	in %	total	urban abs.	in %	total	urban abs.	in %
Bulgaria	8,518	4,383	51.0	8,860	5,051	57.0	9,173	5,724	62.4	9,427	6,365	67.5
Czechoslovakia	14,681	7,647	52.1	15,245	8,308	54.5	15,772	8,972	56.9	16,173	9,582	59.2
Yugoslavia	20,573	7,574	36.8	21,708	9,139	42.1	22,834	10,867	47.6	23,848	12,678	53.2
Hungary	10,297	4,823	46.8	10,528	5,127	48.7	10,793	5,457	50.6	11,010	5,772	52.4
GDR	17,257	14,072	81.1	17,449	14,427	82.7	17,680	14,864	84.1	17,980	15,350	85.4
Poland	33,019	16,842	51.0	34,727	18,998	54.7	36,557	21,335	58.4	38,248	23,688	61.9
Romania	20,309	8,467	41.9	21,418	9,830	45.9	22,417	11,248	50.2	23,310	12,695	54.5
Soviet Union	242,612	138,568	57.1	255,584	155,179	60.7	270,631	173,756	64.2	286,874	193,780	67.5

Source: Demographic Yearbook, UN, 1973.

Conclusions

3. Planning and developing the entire settlement system of a country, entailing gradual integration of urban and rural settlements. This strategy includes strengthening, or establishing, strong regional centers that should stimulate economic, social, and cultural development in the areas.

4. Eliminating differences between town and country, both by means of what is termed indirect urbanization, that is, diffusing the civilization qualities of town life to rural communities, and providing wider job opportunities than in agriculture alone, as well as making the important components of public facilities more easily available.

5. Speeding the process of concentrating rural settlement by restricting the growth in size and number of the smallest communities. This strategy is also one of the effective long-term means for eliminating differences between town and country.

6. A strategy for slowing the growth of large agglomerations is linked with more rational spatial distribution of production, housing, and transportation within these agglomerations, so that the economic advantages of concentration are maintained while stimultaneously improving the social and ecological conditions in the area concerned.

7. Ensuring good environments as far as possible in all parts of settlement, including industrial agglomerations and cities.

Specific and Variable Features

The countries we have been considering started the socialist transformation of their societies in different phases of their economic and settlement development. Some entered the socialist period with relatively advanced industrial bases that had originated under capitalism and had, in some cases, also become concentrated in large agglomerations. Others, while having inherited strong industrial centers or agglomerations, also had along with them backward areas where industry was almost entirely lacking and urban settlement was sparse. In a number

Urbanization in Socialist Countries

of socialist countries, however, urbanization coincided almost completely with the process of socialist industrialization, which created entirely new industrial bases where older, sporadically occurring industry played no important part. This was the case in Bulgaria, Romania, and Yugoslavia. Special and quite varied conditions in relation to levels of industrialization and urbanization existed, of course, in the Soviet Union at the time when it started its rapid industrialization.

In addition to these differences in the industrialization process, which were accentuated by divergences in the sectoral structures of their industries, the socialist countries had, and still have, differing settlement structures and, especially in the past, differing demographic conditions. Consequently, in the first postwar decades the urbanization process followed a somewhat different course in each country, producing settlement structures which, although they have much in common, also possess many specific features. Also significant as a factor causing differentiation in urbanization policies is the size of each country and its macroregional structure.

Under these circumstances each of the general principles for developing settlement in the socialist society underwent some adaptations to prevailing conditions and was used in a way suited to the situation in the given country. It can also be said that in some countries greater weight was attributed to some general principles and less to others. Moreover, principles are fairly numerous; some complement each other, but others require searching for a proper balance that depends on many circumstances, traditions, and public preferences. It is natural, then, that the strategies for developing settlement systems are not identical in all the studied countries, and that from the standpoint of timing their stages as well, they place varying emphasis on this or that component and goal.

The specific approaches are evident, for instance, in applying the concepts of "even distribution in settlement development" and "eliminating differences between regions." The very idea of even distribution is open to various interpretations that can be found in the literature. For instance, population density

Conclusions

can be even, as can the densities of industrial localities, of settlements themselves, or of public facilities (evenly distributed network of schools, etc.). But the concept can also relate to the social dimension of settlement, i.e., equal living standards, equal wage levels, levels of employment, or a uniform employment structure, meaning equal provision of industrial and nonindustrial job opportunities; or it can refer to equal accessibility to the basic public facilities.

The first version, which emphasizes equal densities of population and settlement throughout a country, and which was part of radical decentralizing concepts for settlement development, has been subjected to criticism and has no outstanding supporters today. A. J. Probst, in a critical assessment of the theory that aims, as he says, at "forced equalization without exception," stresses that "...the conditions of life for people and the conditions for production and labor productivity belong in different spheres..."; "...in order to increase the productivity of labor and achieve the maximum level of economic development for a whole country, it is not necessary to attempt maximum development in areas where the conditions for living and for production are inferior."[1]

The second version, relating to the social aspect, is generally accepted, and some writers suggest that it should be separated from views on "economic-industrial" equalization. Probst probably expresses the standpoint of most experts when he says that territorial equalization of living conditions "...must be achieved as rapidly as possible and to the maximum. The population must enjoy a level of material well-being, of cultural and other development that, if not the same in all areas (if that is not fully possible owing to specific natural conditions), should in any case be maximally close."[2]

Demands for reducing regional differences are voiced in all the proposals we have studied, whether explicitly or implicitly. However, there are variations in emphasis. In the smaller countries that have reached higher phases of industrialization, whose urban settlement structures are older, and where the whole territory is at a more advanced stage of economic and social integration, the degree of social equality is fairly great. Examples are the GDR, Czechoslovakia, and in part, Hungary. Owing to relatively high population density, good communications networks, well-developed public and private transportation, and last but not least, thanks also to the large number of dispersed industrial localities, regional differences in living standards, and also in education, health care, and in the accessibility of the basic public facilities, are very slight. Naturally in such cases settlement proposals do not pay as much attention to the problem as in countries where macroregional differences are greater, as in Poland, Yugoslavia, and Romania.

The equalization problem appears in another context in the strongly urbanized countries, that is, in connection with the relation between dispersal and concentration of settlement. The search for a new relation between the two processes is, or has been, an important problem for planners in the socialist countries where the sites of raw materials extraction and inherited industrial capacities were strongly concentrated, for instance, in the GDR and Poland. The extent to which concentration or deconcentration strategies should be applied is also a major issue in countries where the past has left a legacy of macroregional differences in infrastructure levels in areas inhabited by different ethnic groups, which is the case in Yugoslavia and was the case in Czechoslovakia.

Their specific approaches are also evident in the emphasis individual countries place on developing their capital cities and different categories of towns according to size. In this connection Goldzamt's observations are relevant.[3] When comparing the urbanization strategies of the sixties in Poland and Czechoslovakia, he noted that the central problem in the Polish People's Republic was to develop several supraregional cities of several hundred thousand inhabitants that had considerable dynamism, whereas attention had to be centered in Czechoslovakia, in view of its settlement structure, on "topping up" investment in medium-sized and small towns that in the future would become the organizing and key centers of settlement.

Conclusions

Now, however, Czechoslovak urbanization policies tend to stress the formation of settlement agglomerations with several hundreds of thousands of population that should serve as focal points and that have functions analogous to large cities in Poland or the USSR. In Bulgaria and Romania, on the other hand, the accent appears to be more on developing the small and medium-sized towns. In Hungary the area centers are to be strengthened, which means almost all the cities apart from Budapest. In Poland, in accordance with the policy of moderate concentration and polycentrism, the existing urban agglomerations are regarded as the main components in settlement development. The Soviet Union aims to activize its small and its medium-sized towns and recommends slowing down the growth only of cities with more than 500,000 inhabitants.

Attitudes toward capital cities are essentially similar in all the countries, the aim being to reduce growth in size. But the weight given to regulation policy varies. If we can take as representative of the most severely restrictive strategy the Hungarian concept, which aims at selective growth only for Budapest, then the less restrictive policy is that applied in Yugoslavia to Belgrade. The current views on the place of Prague in the Czechoslovak settlement system occupy a more or less middle position.

The countries propose various ways to eliminate differences between town and country according to their settlement densities and other factors. In densely populated areas, where small and medium-sized towns are fairly numerous and good communications make them easily accessible, a substantial number of formerly agricultural communities have been transformed. Places inhabited mainly by farming people become communities with mixed socioeconomic population structures, including increasing numbers of industrial employees. This gives rise to a new social category of nonagricultural rural population. In addition, the proximity of facility centers converts many former rural communities into parts of "indirectly urbanized" areas. This course is followed to a large extent in Czechoslovakia, the GDR, Poland, and Hungary. Stated simply, the rural communities either become parts of larger urbanized territories, or they are totally linked by communications with the nearest town. The prerequisite here is the existence of a relatively dense network of towns distributed over the whole country.

Where the density of settlement and towns is low and daily travel to work or to better-equipped facility centers is hard to envisage, greater emphasis has to be placed on concentrating rural settlement, eliminating small and isolated communities, and creating larger places based on industrial-agricultural production undertakings. This course is recommended in the Soviet Union for areas that will not be included in the group settlement systems that have towns as their cores. They will be in the nature of smaller group systems of a rural type that will serve as instruments in eliminating differences between town and country. Similar proposals have been made in Yugoslavia and the GDR. And some ideas for solving this problem in Romania and Bulgaria are based on cooperation among settlements, a concept that is, in fact, similar to the Soviet group system theory.

No less interesting are trends in the general principles of the strategies. Although the period of ten to fifteen years separating the origins of the first- and second-generation of urbanization and settlement strategies is rather short, it is nevertheless possible to observe that in the period between 1960-65 and 1975 a shift took place in the views on settlement policies.

The main drift of these changes can be summarized as follows:

Greater emphasis is placed on the economic role of the settlement structures in general, this being based on the premise that it is possible for a structure to contribute to greater efficiency in the economy of the whole country.

Rational concentration of economic and social activities is in the majority of the countries studied increasingly regarded as an instrument for achieving such efficiency, but concentration of population and housing is not necessarily included.

Conclusions

An increasingly important role in urbanization strategies is played by the planned formation and regulation of large urban agglomerations, in some cases of continuous urbanized regions. In this context the idea of nodal belt development, leading to the formation of "development axes," acquires greater importance. There is greater tolerance for specific forms in future settlement development that correspond to the economic, social, and settlement conditions in given countries and also in regions within a country — a move away from schematism.

It is also possible to observe less insistence on blueprints for the spatial forms of settlement; the accent is on development in more than one form, be it belt, concentric, or nodal.

In most countries views are changing with regard to desirable levels of concentration, sizes of settlements at different levels in the hierarchy, and the degree of urbanization, while the lower limits for sizes of towns in different categories are being raised.

On the whole, most of the socialist countries stress the need to continue concentrating social activities, and to some extent population as well, in selected settlement units and systems, for instance, regional agglomerations. The principle here is to stimulate concentration not only in individual towns but in larger urbanized areas on the lines of the concept termed "decentralized concentration."

Criticism of the view that the central-place theory represented the only basis on which to work led to the view that the centrality principle should be elaborated and given greater precision; it is no longer seen solely in the traditional form derived from the provision of services for areas surrounding different types of centers. In any case, the schematic versions of centrality, as applied both to rural areas and to areas of concentrated industrial activity, have now been rejected.

There is a search for new forms of balance between the economic, social, and ecological effects of urbanization, which are in many cases conflicting, and the harmonizing of which stimulates the effort to find new organizational and spatial forms for urbanization.

Of growing importance are concepts stressing the complex hierarchical nature of the settlement system, comprising subsystems composed of individual towns and smaller communities that interact intensively. These relatively closed subsystems can be developed only as entities and not by means of plans for single quasi-autonomous towns or communities.

One can observe an ever greater understanding of factors that limit urbanization, especially in countries that are trying to stimulate concentration. In some cases these factors are low population growth; in others they are limits on basic resources such as land and water; and in some, ecological limits.

The dynamic character of settlement systems is now understood better than in the past, and long-term trends in urbanization, together with their macroregional variations, are being looked for. This concern with historical analysis has led, especially in the most recent period, to several socialist countries considering so-called posturbanization phases in the development of their settlement networks.

NOTES

Chapter 1

1. The high degree of urbanization in the ČSSR, and especially the ČSR, is stressed by Miroslav Blažek who, in his paper "Les perspectives de l'urbanization en Tchécoslovaquie" (Blázek 1975, pp. 107-11), compares Czechoslovak urbanization with that in the German Federal Republic (in 1969, 63% of the ČSSR population lived in urban communities, in Germany in 1964, 77.5%). Some other statistical analyses put the ČSSR level lower. The UN Demographic Yearbook for 1973 gives the 1970 figure for the ČSSR as 52.1% living in urban communities. K. Davis, in World Urbanization 1950-1970 (published in 1969), estimates the proportion as 52%. The differences are caused by differing definition of urban communities.

2. J. Mareš, "Změny v rozmístění čs. průmyslu v letech 1930-1960," in Dlouhodobé změny v rozmístění čs. průmyslu, Prague, 1969, p. 33.

3. The data on the proportion of the industrially employed population do not in themselves indicate that all the towns where the figures increased were undergoing industrialization. An increase could be due to more jobs being available in the small towns, but it could also be that more people are commuting from small towns to work in industries commonly situated in larger towns. In the sociological sense, however, the figures demonstrate a penetration by industry into the rural areas and small towns.

4. M. Střída, discussion paper in Dlouhodobé změny v rozmístění čs. průmyslu, no. 20, Department of Economic History series, School of Economics, Prague, 1969.

5. The trend in the first postwar years of ideas on the distribution of industry has been summarized by Dan Gawrecki in his paper "Hlavní tendence v rozmísťování průmyslu v Československu po roce 1945," in Dlouhodobé změny v rozmístění čs. průmyslu.

6. Návrh zásad koncepcí osídlení v ČSSR, Prague, VÚVA, 1964; see also the VÚVA collection, Investice a životní prostředí, Prague, VÚVA, 1966; Rozvoj životního prostředí měst a vesnic v ČSSR, Brno, VÚVA, 1966.

7. Prague, VÚVA, 1963.

Notes to Pages 30-45

8. Principles for Realizing the Long-range Development of Settlement in the Czech Socialist Republic, adopted by the government as Decision No. 283 in 1971.
9. Etarea, Studie životního prostředí, Prague, SKT-PPÚ, 1967.
10. Z. Lakomý, Civilizace, kultura, životní prostředí, Prague, ČSAV, 1972.
11. Law No. 50 was actually the third version of the legal norm regulating physical planning after World War II. In 1949 Law No. 280 on physical planning and the building of communities was promulgated, stressing for the first time the role of physical planning. It was amended in 1958, and for the first time the new version legalized the idea of physical plans for "raions," i.e., large areas constituting economically or culturally important entities (e.g., industrial agglomerations).
12. Terplan is the state institute for regional planning; its work is aimed at drafting plans for large regions.
13. See the Terplan work, Koncepce vývoje osídlení a urbanizace ČSR 1970-2000, Prague, 1975, pp. 14-15.
14. Koncepce vývoje osídlení a urbanizace ČSR 1970-2000, Prague, 1975, p. 9.
15. Basic urbanization is the proportion of population in communities of 5,000 and over; medium denotes the proportion in communities of 20,000 and over; and metropolitan, in those with 100,000 and over.
16. For urbanization development, see P. Zibrín, "Zhodnotenie doteraj síeho vývoja urbanizácie na Slovensku, dosiahnuté výsledky a perspektivy rozvoje," in the collection Národné sympozium, Koncepcia a metodické postupy hlavných smerov urbanizácie na Slovensku, Bratislava, 1971; L. Faltan, Urbanizacja Slowacji po drugiej wojnie światowej, Warsaw, 1975.
17. See A. Mrázik, "Celoslovenské riešenie projektu urbanizácie SSR," Urbanita, 1974, no. 3, pp. 38-44.
18. For the demands made on SSR urbanization, see I. Michalec, "Základné principy koncepcie urbanizácie Slovenska," Životné prostredie, 1973, no. 4, pp. 184-88.
19. Ibid., p. 186.
20. This summary of the principles is based primarily on the Michalec article, and on "Projekt ubranizácie Slovenskej socialistickej republiky," Investiční výstavba, 1974, no. 9, pp. 297-304; "Projekt urbanizácie SSR," Hospodářské noviny, 1975, no. 12, p. 7; P. Zibrin, "Výhl'ady rovoja urbanizácie na Slovensku," Sociologia, 1976, no. 2, pp. 99-158.
21. In the terminology of the project, "urban region" denotes the highest level of spatial settlement structure and consists of the most intensely urbanized complex of activities in an urbanization area. "Urbanization locality" denotes a lower-level but homogeneous spatial settlement structure delimited around settlements with economic bases of a kind to which at least 1,000 people come daily to work (see A. Mrázik, op. cit., p. 40).

Chapter 2

1. See M. L. Strogina, Sotsial'no-ekonomicheskie problemy razvitiia bol'shikh

gorodov v SSSR, Moscow, 1970; F. M. Listengurt, "Dinamika krupneishikh gorodov SSSR," Geografiia, 1970, no. 3.
2. See Iu. L. Pivovarov, Sovremennaia urbanizatsiia, Moscow, 1976, p. 140.
3. B. S. Khorev, Problemy gorodov, Moscow, 1975, chap. 7, "Changes in the Distribution of Settlement and the Zoning of Settlement in the USSR," pp. 237-46.
4. I. P. Muravev and S. V. Uspenskii, Metodologicheskie problemy planirovaniia gorodskogo rasseleniia pri sotsializme, Leningrad, 1974, p. 50.
5. Ibid., pp. 55-56.
6. Ibid., pp. 38-39.
7. Chauncy D. Harris starts his book Cities of the Soviet Union (Chicago, 1970) with the words: "The Soviet Union is a land of cities."
8. V. I. Lenin, Selected Works, vol. 2, Moscow, Progress Publishers, 1967, p. 684.
9. E. Goldzamt, Urbanistyka krajów socjalistycznych, Warsaw, 1971, p. 42.
10. These studies follow the urban belt theory, advanced in a form corresponding to the economic and social conditions obtaining in 1929 in the socialist society of the USSR, by N. A. Miliutin in his book Problems of the Construction of Socialist Towns. Basic Questions of Rational Planning and Settlement Building in the USSR, which was published in Moscow and Leningrad in 1930.
11. See Stroitel'naia gazeta, October 10, 1965.
12. V. G. Davidovich, "O razvitii seti gorodov SSSR za 40 let," Voprosy geografii, 1959, no. 45, pp. 67-68; A. E. Probst, Voprosy razmeshcheniia sotsialisticheskoi promyshlennosti, Moscow, 1971, p. 108; Raionnaia planirovka ekonomicheskikh administrativnykh raionov promyshlenykh raionov i uzlov, Moscow, 1962, p. 62; I. Bocharov, "K probleme optimal'nogo goroda," Arkhitektura SSSR, no. 12, 1960; A. Skvortsov, "Nazrevshie voprosy organizatsii i planirovaniia gorodskogo khoziaistva SSSR," Voprosy ekonomiki, no. 4, 1958; K. F. Kniazev, "K probleme optimal'nogo velichiny i struktury novykh gorodov," Problemy sovetskogo gradostroitel'stva, no. 13, 1960.
13. V. S. Riazanov, "Problemy preobrazovaniia sel," Arkhitektura SSSR, no. 12, 1971.
14. Ibid.
15. B. Svetlichnyi, "Razmyshlaia o sudbakh gorodov," Arkhitektura SSSR, 1967, no. 4.
16. See V. Shkvarikov, "Problemy rasseleniia na sovremennom etape," Arkhitektura SSSR, 1970, no. 9, pp. 12-14; N. Baranov, "O budushchem razvitii sovetskogo gradostroitel'stva," Arkhitektura SSSR, 1970, no. 10, pp. 12-14.
17. B. Belousov, "Osnovnye problemy sovershenstvovaniia sistemy rasseleniia," Arkhitektura SSSR, 1974, no. 3, pp. 3-12.
18. See P. Vladimirov and A. Kochetkov, "Problemy perspektivnogo rasseleniia," Arkhitektura SSSR, 1974, no. 5, pp. 2-3.
19. The term "scheme" denotes a general concept or strategy for development.
20. G. Kaplan, A. Kochetkov, and F. Listengurt, "Zur Umgestaltung des Siedlungsnetzes und zum Aufbau von Gruppensiedlungen," Deutsche Architektur, 1972, no. 2, pp. 113-14.
21. G. N. Fomin, "Sistemy rasseleniia SSSR kak ob"ekty prognozirovaniia,"

Notes to Pages 72-93

in Gradostroitel'stvo, Kiev, 1974, p. 17.
22. B. S. Khorev and S. Smidovich, "The Settlement System of the Russian Soviet Federative Socialist Republic," International Geographical Union, Commission on National Settlement Systems, duplicated address to a session of the IGY in 1979 in Szymbark (Poland).
23. See G. N. Fomin, "Sovetskoe gradostroitel'stvo na novom etape," Kommunist, 1974, no. 11, p. 47.
24. O. S. Pchelintsev, "Urbanizatsiia, regional'noe razvitie i nauchnotekhnicheskaia revoliutsiia," Ekonomika i matematicheskie metody, 1978, vol. XIV, no. 1, pp. 6-20.
25. Ibid., p. 7.

Chapter 3

1. J. Dangel, Przekształcenia sieci mieskiej w Polsce pod wpływem rozwoju ludności i uprzemysłowienia kraju w okresie 1946-1960, Warsaw, 1968.
2. Ibid., p. 214.
3. A. Stasiak, "Specyfika polskiej drogi urbanizacji w świetle wyników NSP," Miasto, 1971, no. 6, p. 9.
4. R. Grabowiecki, "Plan przestrzenny Polski do roku 1990," Gospodarka planowa, 1974, no. 4, p. 215.
5. E. Goldzamt, Urbanistyka krajów socjalistycznych, Warsaw, 1971, p. 20.
6. See Antoni Kukliński and Michał Najgrakowski, "Industrializacja, urbanizacja i rozwój regionalny w Polsce," Miasto, 1979, no. 1-2, pp. 5-16.
7. Ibid., p. 15.
8. Our account of the origin of nationwide regional planning is primarily based on Boleslaw Malisz's book Problematyka przestrzennego zagospodarowania kraju, Warsaw, 1974; see especially pp. 11-27.
9. In addition to the universities, which have physical planning institutes and specialized research institutes for urban and area planning, a notable feature of the research apparatus in Poland is the fact that the Academy of Sciences deals intentively with settlement questions, not only in the Institute of Geography but also through the Commission for the National Development Plan. There are also many regional research institutes concerned with settlement in their own provinces, e.g., the Silesian Institute of Sciences in Katowice.
10. T. Mrzygłód, Zalozenia przestrzennego zagospodarowania kraju w okresie perspektywniczym do 1985 r., Warsaw, 1968.
11. B. Malisz, "Teoretyczny model sieci miast w Polsce," in Problemy rozwoju sieci osadniczej, Warsaw, 1967.
12. B. Malisz, Zarys teorii kszaltowania układów osadniczych, Warsaw, 1966.
13. See Rozwój spoleczny Polski w pracach prognostycznych, Warsaw, PAN, Komitet "Polska 2000," 1971.
14. R. Grabowiecki, "Plan przestrzenny Polski do roku 1990," Gospodarka planowa, 1974, no. 4, pp. 214-21.
15. For instance, concepts by B. Malisz and P. Zaremba were joined in a sin-

Notes to Pages 93-107

gle proposal contained in a publication by the two authors in 1971. Similarly, concepts advanced by S. Leszczycki and B. Malisz were merged to become one of the bases for the official strategy produced by the Planning Commission.

16. S. Leszczycki, P. Eberhardt, and S. Herman, "Prognoza przestrzennogo zagospodarowania kraju do roku 2000," in Rozwój społeczny Polski w pracach prognostycznych, Warsaw, PAN, Komitet "Polska 2000," 1971.

17. R. Karlowicz, "System wielkich aglomeracji mieskich w Polsce," Miasto, 1972, no. 12, pp. 1-10.

18. K. Dziewoński, "Hipoteza przekształceń sieci osadniczej w Polski do roku 2000," in Polska 2000. Prognozy rozwoju sieci osadniczej, PAN, 1971, no. 2, pp. 96-107.

19. B. Malisz, "Prognóza zmian sieci osadniczej w Polsce," in ibid., pp. 248-50; B. Malisz, Problematyka przestrzennego zagospodarowania kraju, Warsaw, 1974, chap. 4, pp. 98-137.

20. P. Zaremba, "Proba prognozy rozwoju śieci osadniczej w Polsce," in ibid., PAN, no. 2., pp. 108-47.

21. "Ewolucja struktury przestrzennej kraju," in Procesy urbanizacji kraju w okresie XXX-lecia Polskiej Rzeczypospolitej Ludowej, Jan Turowski, Warsaw, PAN, 1978, pp. 49-50.

Chapter 4

1. D. Scholz, "Die wirtschaftsräumliche Struktur der DDR," Geographische Berichte, 1971, no. 2.

2. H. Lüdeman and J. Heinzmann state in their paper "The Interrelations between the Process of Concentration and Territorial Development under Socialist Conditions in the GDR" (IGU Conference, Regional Studies, Methods and Analyses, Budapest, 1974) that the proportion of the population in these towns rose from 12.7% in 1952 to 15.3% in 1969.

3. A. von Känel and D. Scholz, "Wirtschaftsräumliche Struktureinheiten mittlerer Ordnung in der DDR," Petermanns Geogr. Mitteilungen, vol. 3, 1969.

4. For more detail about this distinction, see Frankdieter Grimm, The Settlement System of the German Democratic Republic, Report to the Commission on National Settlement Systems of the International Geographical Union, Institut für Geographie und Geoökologie der Akademie der Wissenschaft der DDR, Leipzig, 1979, pp. 5-7.

5. Details on the course and conditions of urbanization in the GDR are to be found in a paper by Konrad Scherf, "Zu einigen ökonomischen, sozialen und territorialen Aspekten der sozialistischen Urbanisierung in der DDR," Wirtschaftswissenschaft, 1979, no. 1, pp. 31-46.

6. Ibid., p. 39.

7. This view has been expressed, for instance, by Horst Behr in his paper "Die Wirtschaftskraft eines Ballungsgebiete im Vergleich zum Territorium der DDR — eine Analyse ausgewählter Kennziffern," Wirtschaftswissenschaft, 1979, no. 1, pp. 47-60.

8. H. Lüdemann and J. Heinzmann, "The Interrelation between the Process of Concentration and Territorial Development under Socialist Conditions in the GDR," IGU Conference, Regional Studies, Methods and Analyses, Budapest, 1974, p. 318.
9. W. Ostwald, "Planmässige proportionale Gestaltung der Territorialstruktur der DDR als Beitrag zur Erhöhung der effektivität der gesellschaftlichen Reproduktion," Wirtschaftswissenschaft, 1976, no. 5, pp. 656-73.
10. Ibid., p. 316.
11. See F. Grimm, op. cit., pp. 5-7.
12. F. Grimm and I. Hönsch, "Zur Typisierung der Zentren in der DDR nach ihrer Umlandbedeutung," Petermanns Geogr. Mitteilungen, vol. 118, no. 4, pp. 282-88.
13. G. Mohs, R. Schmidt, and D. Scholz, "Territoriale Konzentration und Urbanisierung," Petermanns Geogr. Mitteilungen, vol. 120, no. 2, 1976, p. 90.

Chapter 5

1. See E. Lettrich, "Urbanizálódás Magyarországon," Budapest, 1965; the abstract Urbanization in Hungary, 1967, p. 8.
2. P. Beluszky, A magyar városok központi szerepköre; "Izmeneniia nastupivshie v strukture seti naselennykh tsentrov v Vengrii," Varna, 1971 (manuscript of a lecture).
3. I. Bencze, "The Role of Capital Cities in Socio-economic Development," in IGU, Regional Studies, Methods and Analyses, Budapest 1974, pp. 231-49.
4. György Enyedi points to this in his paper "Regional Development Policy in Hungary" (in manuscript).
5. E. Lettrich, Urbanizálódás Magyarországon, Budapest, 1965, p. 18.
6. "Rozvoj sídel and sídlištní sítě v Maďarsku" (Czech translantion of an official document presented by the Hungarian delegation to the UN "Habitat" conference on settlement, Vancouver, 1976), Research Institute of Building and Architecture, Brno, 1975.
7. L. Fodor and I. Illés, "Metropolitan Industrial Agglomeration," Regional Science Association, Papers, vol. XXII, 1968; V. Tajti-Erzsébet, Budapest munkaerővonzása.
8. On the specific problems of Hungarian settlement, see K. Perczel, "Les facteurs de l'urbanisation et de concentration de la population en Hongrie," and E. Lettrich, "Les traits caractéristiques de l'urbanisation en Hongrie," in IGU, Urbanization in Europe, Budapest 1975. Also see K. Mihailović's view in his book Regional Development. Experiences and Prospects in Eastern Europe, The Hague, 1973, p. 88, where he writes that "these enormous villages cannot as yet be industrialized. Urbanization appears to be the best solution for the future for these villages in Hungary and in the Yugoslav Vojvodina." This is a long-term program, to be implemented gradually.
9. A detailed examination of the study and the preparatory work for it is

Notes to Pages 122-38

given by V. Matoušek in his paper "Hlavní směry přestavby osídlení v evropských zemích a jejich srovnání se zásadami přestavby sídlištní sítě ČSSR," Brno, VÚVA, 1967; also see G. Köszegfalvi, "Einige Probleme der künftigen Entwicklung des Siedlungsnetzes in der Ungarischen Volksrepublik," Deutsche Architektur, 1964, no. 10.
   10. L. Fodor and I. Illés, "Metropolitan Industrial Agglomeration," Regional Science Association, Papers, vol. XXII, 1968, pp. 67-72.
   11. K. Perczel, "Budapest Túlsúfoltságának hatása az országos. Településhálozat fejlessztésére," Épitesügyi Szemle, 1963, no. 2.
   12. K. Perczel, "A városközpontok országos rendszere és munkamegosztása," Területrendezés, 1972, no. 3, pp. 61-73.
   13. Intensive economic development, i.e., reproduction, is understood in socialist economics to signify raising output by more efficient utilization of productive resources. Investment, in this case, is directed to improving the technological base and equipping labor with more efficient machinery. Performance is improved with a constant or declining work force.
   14. "Rozvoj sídel a sídlištní sítě v Mad'arsku," p. 8.
   15. See ibid.

Chapter 6

1. See C. Rusenescu, "L'urbanisation et les nouveaux rapports entre la population, le site et le territoire de villages roumains," in IGU, Urbanization in Europe, Budapest, 1975.
   2. See V. Trebici, "Reflectii demografie în legătură cu prognoza populatici urbane," Architektura (Bucharest), 1972, no. 3-4, pp. 25-26.
   3. V. Matoušek, Hlavni směry přestavby osídlení v evropských zemích a jejich srovnání se zásadami přestavby sídlištní sítě CSSR, Brno, VÚVA, 1967.
   4. P. Deica and J. Stefǎnescu, "Forms of the Territorial Grouping of the Settlement Network in the Socialist Republic of Romania," Revue roumaine de géologie, géophysique et géographie. Serie de géographie, 1972, no. 2.
   5. Our review of settlement development in Bulgaria relies, in particular, on the following works: N. Michev, "L'urbanization et le probleme du développement du réseau urbain en Bulgarie," in IGU, Urbanization in Europe, Budapest, 1975, pp. 99-106; P. Grekov, "Saobrazheniia za tendentsiite v razvitieto na gradoustroistvoto," Izvestiia na nauchnoizsledovatelskiia institut po gradoustroistvo i arkhitektura, book XXIII, vol. 5, Sofia, 1971; L. Tonev, "Funktsionalna organizatsiia i struktura na selishtnite edintsi, grupi i sistemi," Arkhitektura (Sofia), 1973, no. 2.
   6. P. Grekov, op. cit.
   7. G. Sebestyén, "Některé předběžné předpoklady a hypotézy prognozy procesu urbanizace v RSR," Architektura (Prague), 1972, no. 3-4.
   8. G. Gusti, "Náměty na uspořádání struktury osídlení v Rumunsku," Architektura (Prague), 1968, no. 4.
   9. Quoted from an article on the Tenth Congress of the Bulgarian Communist

Party in Arkhitektura (Sofia), 1973, no. 1.
10. L. Tonev, op. cit.
11. P. Grekov, op. cit.
12. C. Cănilă, "Aspecte ale urbanizării in R. S. România," Terra,1974, no. 4.
13. Problemi na geografiiata na naselenieto i selishchata, Sofia, 1973, p. 191.

Chapter 7

1. I. Ginić, Dinamika i struktura gradskog stanovištva Jugoslavije, Belgrade, 1967.
2. D. Štefanović and S. Žuljić, Ekonomski aspekti izgradnje gradova u Jugoslavii, Belgrade, 1966.
3. The Hague, 1973, p. 83.
4. M. Rančić, "Značenije i problemy demografičeskich prognozov naselenych punktov v Jugoslavii," International Conference of Demographers of the Socialist Countries, Liblice, 1976.
5. Described according to information in V. Matoušek's study of the main lines of settlement reconstruction and Z. Ryšavý's report on his visit to Yugoslavia in 1973.
6. I. Vrišer, "Städte in Slovenien," in Sozialgeographische Probleme Südosteuropas, K. Ruppert, ed., Regensburg, 1973.
7. Vladimir Kokole, "Osnove policentričnega urbanega sistema v SR Sloveniji," Zavod SRS za regionalno prostorsko planiranje, no. 29, Ljubljana, 1975.
8. Zdravko Mlinar, "Urban Growth and Social Development," paper presented at the Ninth World Congress of Sociology, Uppsala, 1978.

Chapter 8

1. A. J. Probst, Voprosy razmeshcheniia sotsialisticheskoi promyshlennosti, Moscow, 1971, pp. 64-65.
2. Ibid., pp. 67-68.
3. E. Goldzamt, Urbanistyka krajów socjalistycznych, Warsaw, 1971, p. 102.

SELECTED BIBLIOGRAPHY

General Problems of Urbanization

Akhiezer, A. S. "Nauchno-tekhnicheskaia revoliuciia i upravlenie razvitiem obshchestva," Voprosy filosofii, 1968, no. 8.
Akhiezer, A. S.; Kogan, L. B.; and Ianitskii, O. N. "Urbanizatsiia, obshchestvo i nauchno-tekhnicheskaia revoliutsiia," Voprosy filosofii, 1969, no. 2.
Berry, B. J. L. The Human Consequences of Urbanization (particularly "The City of Socialist Man," pp. 154-63, and chap. 5, "Divergent Paths in Twentieth-Century Urbanization," pp. 164-81), London, 1973.
Bonifateva, L. I., and Pokshishevskii, V. V. "Urbanizatsiia," Kratkaia geograficheskaia entsiklopediia, vol. 5, Moscow, 1966.
Hampl, M. Teorie komplexity a diferenciace světa, Prague, 1971.
Ianitskii, O. N. "Sotsial'no-informatsionnye aspekty urbanizatsii," in Urbanizatsiia i rabochii klass v usloviiakh nauchno-tekhnicheskoi revoliutsiia, Moscow, 1970.
———. "Sotsial'nye aspekty urbanizatsii v usloviiakh nauchno-tekhnicheskoi revoliutsii," in Iu. L. Pivovarov, ed., Problemy sovremennoi urbanizatsii, Moscow, 1972.
———. Urbanizatsiia i sotsial'nye protivorechiia kapitalizma, Moscow, 1975.
Ivanov, V. "Urbanizatsiia: nastoiashchee i budushchee," Voprosy ekonomiki, 1969, no. 11.
Jachiel, N. Gradat i seloto, Sociologičeski aspekti, Sofia, 1965.
Jałowiecki, B. Miasto i społeczne problemy urbanizacji, Warsaw, 1972.
Khorev, B. S. Problemy gorodov, Moscow, 1st ed., 1972; 2nd ed., 1975.
Kogan, L. B. "Urbanizatsiia," Filosofskaia entsiklopediia, vol. 5, Moscow, 1970.
Mlinar, Z. "Urban Growth and Social Development," Ninth World Congress of Sociology, WG 13, Rural and Urban Development, Uppsala, 1978.
Musil, J. Sociologie soudobého města, Prague, 1967.
———. Urbanizace v socialistických zemích, Prague, 1977.
Pivovarov, Iu. L. "Urbanizatsiia i nauchno-tekhnicheskaia revoliutsiia," Izvestiia AN SSSR, Seriia geograficheskaia, 1970, no. 5.
———. "Sovremennaia urbanizatsiia: sushchnost', faktory i osobennosti

Urbanization in Socialist Countries

izucheniia," in Iu. L. Pivovarov, ed., Problemy sovremennoi urbanizatsii, Moscow, 1972.

_____. "Kontsentratsiia funktsii i tendentsii urbanizirovannogo rasseleniia," in Resursy, sreda, rasselenie, Moscow, 1974.

_____. Sovremennaia urbanizatsiia, Moscow, 1976.

Purš, J. Průmyslová revoluce, Prague, 1973.

Röhr, F. "Inhalt und Tendenzen der Urbanisierung," Architektur der DDR, vol. XXV, no. 11, 1976, pp. 665-69.

Rybicki, P. Społeczeństwo miejskie, Warsaw, 1972.

Urbanizatsiia i rabochii klass v usloviiakh nauchno-tekhnicheskoi revoliutsii, Moscow, 1970.

"Urbanizatsiia mira," Voprosy geografii, no. 96, Moscow, 1974.

Urbańska, B. "Typy społeczności urbanizujacych sie a typy uprzemysłowania," Górnośląskie studia socjologiczne, 1967, no. 7.

Ziółkowski, J. Urbanizacja, miasto, osiedle, Warsaw, 1965.

Economic and Planning Problems in Urbanization

Akhiezer, A. S., and Kochetkov, A. V. "Urbanizatsiia i intensifikatsiia proizvodstva v SSSR," in Iu. L. Pivovarov, ed., Problemy sovremennoi urbanizatsii, Moscow, 1972.

Artemchuk, V. I. Metodika opredeleniia stoimosti stroitel'stva i eksploatatsii gorodov, Kiev, 1964.

Bocharov, I.; Markus, B.; and Simbirtsev, V. "Opyt proektirovaniia optimal'nogo goroda," Arkhitektura SSSR, 1960, no. 12.

Davidovich, V. G. "O razvitii seti gorodov SSSR za 40 let," Voprosy geografii, 1959.

Fedotova, N. V. "Ob ispol'zovanii pokazatelia fondootdachi dlia analiza razmeshcheniia promyshlennogo proizvodstva," in Problemy razvitiia gorodov (no date).

Gałeski, B. "Typy uprzemysłowienia," Studia socjologiczne, 1967, no. 4.

Ginsbert, A. "Ekonomiczne przesłanki rozwoju malych miast," Miasto, 1964, no. 10.

Hegedüs, M. "Adalékok a hazai urbanizáció megítéléséhez," Városépités, 1972, no. 3.

Ivković-Ivandekič, P. "Urbanizacja a wzrost gospodarczy," Miasto, 1962, no. 12.

Kasalický, V. The Future of Urban Evironment, London, 1972.

Klacek, J., and Klacková, J. "Úloha produktivity práce v urbanizačním procesu," Politická ekonomie, vol. XXVI, no. 5, 1978, pp. 407-17.

Kniazev, K. F. "K probleme optimal'noi velichiny i struktury novykh gorodov," Problemy sovetskogo gradostroitel'stva, 1960, no. 13.

Kozlovskaia, L. V. "K voprosu ob efektivnosti territorial'noi kontsentratsii promyshlennosti," Materialy Moskovskogo filiala Geograficheskogo obshchestva SSSR, Seriia Geografiia naseleniia, vol. 3, 1969.

_____. Territorial'naia kontsentratsiia promyshlennosti, Minsk, 1975.

Selected Bibliography

Kukliński, A. "Regiony silne i slabe w polityce społeczno-ekonomicznej," Przegląd Geograficzny, vol. XLVIII, no. 3, 1976, pp. 389-400.

_____. "Investycje, urbanizacja i efektywność rozwoju regionalnego. Problemy dyskusyjne," in Procesy urbanizacji kraju w okresie XXX-lecia Polskiej Rzeczypospolitej Ludowej, J. Turowski, ed., Warsaw, 1978, pp. 57-66.

Lappo, G. M., and Pivovarov, Iu. L. "Urbanizatsiia i prirodnaia sreda," in Geograficheskie aspekty urbanizatsii, Moscow, 1971.

Ledworowski, B. "Niektore aspekty ekonomiczne rozwoju sieci osadniczej," Miasto, 1968, no. 3.

Leszczycki, St. "Metody aktywizacji obszarów slabiej rozwinietych," Przegląd Geograficzny, vol. XLVIII, no. 3, 1976, pp. 379-88.

Líkař, O.; Musil, J.; et al. Metodika hodnocení ekonomické a sociální efektivnosti změn v osídlení, Prague, VÚVA, 1975.

Malisz, B. "Teorja Progów," Biuletin Instytutu Urbanistyki i Architektury, 1963, no. 16-17.

Mihailovič, K. Regional Development. Experiences and Prospects in Eastern Europe, The Hague, 1973.

Muravev, I. P., and Uspenskii, S. V. Metodologicheskie problemy planirovaniia gorodskogo rasseleniia pri sotsializme, Leningrad, 1974.

Muszyński, M. Ekonomiczna ocena dwuzawodowości w Polsce, Warsaw, PWN, 1973.

Nekrasov, N. N. Regional'naia ekonomika, Moscow, 1975.

Nekrasov, N. N., and Kormnov, Iu. F. Regional'nye problemy i territorial'noe planirovaniie v sotsialisticheskikh stranakh Evropy, Moscow, 1976.

Pchelintsev, O. S. "Problemy razvitiia bol'shikh gorodov," in Sotsiologiia v SSSR, vol. 2, Moscow, 1966.

_____. "Urbanizatsiia, regional'noe razvitie i nauchno-tekhnicheskaia revoliutsiia," Ekonomika i matematicheskie metody, vol. XIV, no. 1, 1978, pp. 5-20.

Planowanie rozwoju regionalnego w krajach europejskich, A. Kukliński, ed., Warsaw, 1976.

Probst, A. I. Voprosy razmeshcheniia sotsialisticheskoi promyshlennosti, Moscow, 1971.

Ryšavý, Zd., and Kotačka, L. "Vliv obslužného a výrobního sektoru na sídelní strukturu," Architektura ČSSR, 1976, no. 4.

Secomski, K. Wstęp do teorii rozmieszczenia sił wytwórczych, Warsaw, 1956.

Skvortsov, A. "Nazrevshie voprosy organizatsii i planirovaniia gorodskogo khoziaistva SSSR," Voprosy ekonomiki, 1958, no. 4.

Sokolov, V. "Models Aiding National Settlement Policies in the USSR: a Survey," Environment and Planning, vol. VII, no. 7, 1975, pp. 757-80.

Turowski, J. "Kierunki i rodzaje industrializacji i rozwój społeczno-gospodarczy wsi i rolnictwa," Studia socjologiczne, 1972, no. 4.

Zawadzki, S. M. Podstawy planowania regionalnego, Warsaw, 1969.

Zbořil, M. "Teorie a kritéria hodnocení 'prahů' v rozvoji měst," Architektura ČSSR, 1967, no. 8.

_____. "Proces urbanizace a jeho vliv na rozvoj ekonomiky země," Po-

litická ekonomie, vol. XXIV, no. 8, 1976, pp. 691-704.
Zechowski, Z. O. O niektorych trudnośćiach rozwoju malych miast w rejonie uprzemysłowianym, Warsaw, 1967.
Ziólkowski, J. "Typy organicznej industrializacji w Polsce Ludowej, Casus Poznań," in Socjologiczne problemy industrializacji w Polsce Ludowej, Warsaw, 1967.

Social and Cultural Aspects of Urbanization

Dobrowolska, D. "Social Change in Suburban Villages," in Rural Social Change in Poland, Warsaw, Polish Academy of Sciences, 1976.
Dolgii, V. M.; Levada, I. A.; and Levinson, A. G. "Urbanizatsiia kak sotsiokul'turnyi protsess," in N. N. Baranskii, ed., Urbanizatsiia mira, Moscow, 1974.
Franců, D. "Urbanizácia a mezil'udské vztahy," Sociólogia, 1976, no. 2.
Ianickii, O. N. "Gorod kak informatsionnaia sistema," in Sotsiologicheskie issledovaniia goroda, Informatsionnyi biulleten, 1969, no. 1/16.
Jałowiecki, B. Spoleczne procesy rozwoju miasta (zejména kapitola první Processy Urbanizacji), Katowice, Sląski Instytut Naukowy, 1976.
Jarosz, M. "Zjawiska dezorganizacji społecznej zwiazane z procesami in dustrializacji i urbanizacji," in Procesy urbanizacji kraju w okresie XXX-lecia Polskiej Rzeczypospolitej Ludowej, J. Turowski, ed., Warsaw, 1978, pp. 79-95.
Kagan, M. I. "Urbanizatsiia, prostranstvennaia mobil'nost', podvizhnost'," in Urbanizatsiia i rabochii klass v usloviiakh nauchno-tekhnicheskoi revoliutsii, Moscow, 1970.
Kogan, L. B., and Pravotorova, A. "Sotsial'no-kul'turnye sviazi v aglomeratsii krupneishego goroda i ego razvitie," Arkhitektura SSSR, 1976, no. 1.
Kotačka, L. "Preference sídelních typů mezi obyvatelstvem ČSSR," in Otázky urbanizace, Výzkumný ústav výstavby a architektury, Prague, 1976.
Kowalski, M. "Urbanizacja wsi jako aspekt przemian życia wiejskiego," in Procesy urbanizacji kraju w okresie XXX-lecia Polskiej Rzeczypospolitej Ludowej, J. Turowski, ed., Warsaw, 1978, pp. 267-71.
Librová, E. Vliv urbanizace na některé složky režimu dne, Prague, VÚVA, 1975.
Nowakowski, S. "Modernization as a Result of Urbanization," The Polish Sociological Bulletin, 1973, no. 1-2.
_____. Przemiany miejskich społeczności lokalnych w Polsce, Warsaw, 1974.
Pawelczyńska, A., and Tomaszewska, W. Urbanizacja kultury w Polsce, Warsaw, 1972.
Pietraszek, E. "Uwagi o aspektach i wskaźnikach urbanizacji wsi," in Procesy urbanizacji kraju w okresie XXX-lecia Polskiej Rzeczypospolitej Ludowej, J. Turowski, ed., Warsaw, 1978, pp. 237-46.
Pióro, Z. "Socjologiczna koncepcja rozwoju spolaryzowanego," Roczniki socjologii swi, 1976, vol. 14.

Selected Bibliography

Piotrowski, W. "Socjologiczne problemy urbanizacji wsi," in Procesy urbanizacji kraju w okresie XXX-lecia Polskiej Rzeczypospolitej Ludowej, J. Turowski, ed., Warsaw, 1978, pp. 249-54.

Röhr, F., and Röhr, L. "Urbanisierung und urbane Lebensweise," Architektur der DDR, vol. XXVII, no. 7, 1978, pp. 438-42.

Rybicki, P. "Socjologiczne koncepcje urbanizacji," in Procesy urbanizacji kraju w okresie XXX-lecia Polkiej Rzeczypospolitej Ludowej, J. Turowski, ed., Warsaw, 1978, pp. 21-40.

Sowa, K. "Środowisko społeczne mieszkańca wielkego miasta," Studia socjologiczne, 1971, no. 1.

Szczepański, J. Zmiany spoleczeństwa polskiego w procesie uprzemysłowienia, Warsaw, 1973.

Trasnea, O. "Principales coordonées de la modernisation politique en Roumaine," Revue Roumaine des Sciences Sociales, Serie de Sociologie, 1974, vol. 18.

Turowski, J. "Osiedle im. A. Mickiewicza w Lublinie jako nowe środowisko społeczne w wielkim mieście," in Studi socjologiczne i urbanistyczne miast Lubelszczyzny, Lublin, 1970.

_____. "Socjologiczne aspekty społeczności osiedlowej," Studia socjologiczne, 1973, no. 3.

Tyszka, Zb. "Aktualne badania poznańskiego ośródka socjologicznego dotyczace rodzin wielkomiejskich," in Przemiany społecznosci lokalnych w Polsce, Warsaw, 1974.

Volkov, A. G. "Vliianie urbanizatsii na demograficheskie protsessy v SSSR," in Iu. L. Pivovarov, ed., Problemy sovremennoi urbanizatsii, Moscow, 1972.

Wallis, A. Warszawa i przestrzenny uklad kultury, Warsaw, 1969.

Zagórski, K. "Urbanization and Resulting Changes in Class Structure and Education," International Journal of Sociology, vol. VII, no. 3-4, 1977-78, pp. 48-58.

Ziółkowski, J. Sosnowiec, Drogi i czynniki rozwoju miasta przemyslowego, Katowice, 1960.

_____. "Social Problems of Regional Development," Proceedings of the First Scandinavian-Polish Regional Science Seminar, Warsaw, 1967, pp. 16-32.

Urbanization in Czechoslovakia

Andrle, A., and Pojer, M. "Rozvoj největších měst v ČSSR v letech 1961-1970," Investiční výstavba, 1972, no. 9.

Andrle, A. "Migrace obyvatelstva a růst větších měst ČSSR," Geografický časopis, 1975, no. 3.

_____. "Venkovská sídla v ČSSR," Sborník Československé společnosti zeměpisné, vol. LXXXII, no. 4, 1977, pp. 299-312.

Bína, J. "K tendencím vývoje československé sídelní struktury a středisek," Sborník Československé společnosti zeměpisné, vol. LXXXIII, no. 1, 1978, pp. 29-39.

Blažek, M. Sídla v Československu, Prague, 1951.

Urbanization in Socialist Countries

_____. "Cíle průmyslové oblastní politiky a jejich plnění v letech 1946-1966," in Dlouhodobé změny v rozmístění československého průmyslu, Prague, VŠE, 1969.

_____. "Les perspectives de l'urbanisation en Tchécoslovaquie," in Urbanization in Europe, Budapest, IGU, 1975.

Faltan, L. "Vplyv lokalizacie priemyslu na formovanie sídelnej siete Slovenska v období budovania socialismu," Sociologia, vol. VII, no. 2, 1975, pp. 135-47.

Gawrecki, D. "Hlavní tendence v rozmisťování průmyslu v Československu po roce 1945," in Dlouhodobé změny v rozmístění československého prumyslu, Prague, VŠE, 1969.

Investice a životní prostředí, Prague, VÚVA, 1966.

Jirový, K. "Postup a tempo urbanizace v CSSR se zřetelem k dosavadnímu a příštímu vývoji hlavního města Prahy," Demografie, 1971, no. 1.

Kohout, B., et al. Perspektivní vývoj osídlení ČSR a jeho společenskoekonomické důsledky, Prague, Terplan, 1971.

_____. Prognóza územního uspořádání a urbanizace ČSR, Prague, Terplan, 1972.

_____. Koncepce vývoje osídlení a urbanizace ČSR 1970-2000, Prague, Terplan, 1975.

Koncepce hlavních směrů urbanizace, Prague, Ministerstvo výstavby a techniky ČSR, 1975.

Koubek, J. Urbanizace a koncentrace obyvatelstva evropských socialistických států po druhé světové válce, Prague, VŠE, 1975.

Lakomý, Zd., et al. Civilizace, kultura, životní prostředí, Prague, CSAV, 1972.

Malík, Zd., et al. Urbanizace a rozvoj systému osídlení, Brno, VÚVA, 1975.

Mareš, J. "Změny v rozmístění československého průmyslu v letech 1930-1960," in Dlouhodobé změny v rozmístění československého průmyslu, Prague, VŠE, 1969.

Matoušek, Vl. Hlavní směry přestavby osídlení v evropských zemích a jejich srovnání se zásadami přestavby sídlištní sítě ČSSR, Brno, VÚVA, 1967.

_____. "Hlavní rysy vývoje městského osídlení v evropských zemích," Územní plánování a urbanismus, 1974, no. 5.

_____. "Sídelní regionální aglomerace ČSR — jejich vývoj v rámci osídlení a vztahy k sídelní síti," Výstavba a architektura, vol. XXII, no. 10, 1976, pp. 9-27.

Matoušek, Vl., et al. Rozvoj životního prostředí měst a vesnic v ČSSR, Brno, VÚVA, 1966.

Michalec, I. "Základné principy koncepcie urbanizácie Slovenska," Životné prostredie, 1973, no. 4.

Mrázik, A. "Celoslovenské riešenie projektu urbanizácie SSR," Urbanita, 1974, no. 3.

Návrh zásad koncepcí osídlení v ČSSR, Prague, VÚVA, 1964.

Ryšavý, Zd., and Link, J. "Vliv polohy měst v sídelní síti českých zemí na jejich vývoj," in Otázky urbanizace, Prague, Výzkumný ústav výstavby a architektury, 1976.

Základní otázky osídlení ČSSR, Prague, VÚVA, 1963.

"Zásady k realizaci dlouhodobého vývoje osídlení v České socialistické republice," schválené usnesením vlády ČSR, no. 283, 1971.

Selected Bibliography

Zibrín, P. "Zhodnotenie doterajšieho vývoja urbanizácie na Slovensku, dosiahnuté výsledky a perspektivy rozvoja," in Koncepcia a metodické postupy hlavných smerov urbanizácie na Slovensku, Bratislava, 1971.

Urbanization in the Soviet Union

Baranov, N. "O budushchem razvitiii sovetskogo gradostroitel'stva," Arkhitektura SSSR, 1970, no. 10.

Barchin, M. G. "K probleme rasseleniia," Arkhitektura SSSR, 1967, no. 6.

Belousov, B. "Osnovnye problemy sovershenstvovaniia sistemy rasseleniia," Arkhitektura SSSR, 1974, no. 3.

Fomin, G. N. "Perspektiva formirovaniia gruppovykh sistem naselennykh mest," in Perspektivy preobrazovaniia okruzhaiushchej cheloveka gorodskoi sredy, Moscow, 1973.

―――――. "Sovetskoe gradostroit'elstvo na novom etape," Kommunist, 1974, no. 11, p. 47.

Geograficheskie aspekty urbanizatsii, Moscow, 1971.

Harris, D. Ch. Cities of the Soviet Union, Chicago, 1970.

Kaplan, G.; Kochetkov, A.; and Listengurt, F. "Zur Umgestaltung des Siedlungsnetzes und zum Aufbau von Gruppensiedlungen," Deutsche Architektur, 1972, no. 2.

Lappo, G. M., and Pivovarov, Iu. L. "The Soviet Union," in Essays on World Urbanization, R. Jones, ed., London. 1975.

Larmin, O. V.; Moiseenko, V. M.; and Khorev, B. S. "Sotsial'no-demograficheskie aspekty urbanizatsii v SSSR," in Problemy urbanizatsii v SSSR (no date).

Listengurt, F. M. "Perspektivnye izmeneniia gorodskogo naseleniia v SSSR," Izvestiia AN SSSR, Seriia geograficheskaia, 1969, no. 1.

―――――. "Dinamika krupneishikh gorodov SSSR," Vestnik Moskovskogo universiteta, Geografiia, 1970, no. 3.

―――――. "Osnovnye zakonomernosti protsessa urbanizatsii v SSSR," in Urbanizatsiia mira, Voprosy geografii 96, Moscow, 1974, pp. 98-105.

Maergolts, I. M., and Pivovarov, Iu. L. "Sovremennaia urbanizatsiia v sotsialisticheskich stranakh Evropy," in Urbanizatsia mira, Voprosy geografii 96, Moscow, 1974, pp. 115-23.

Naselenie SSSR 1973, Statisticheskii sbornik, Moscow, 1975.

Planmässige Gestaltung der Siedlungsstruktur und der Urbanisierung in der entwickelten sozialistischen Gesellschaft, Moscow and Berlin, 1978.

"Problemy sovremennoi urbanizatsii," in D. I. Valentei, ed., Marksistsko-leninskaia teoriia narodonaseleniia, Moscow, 1971.

Riazanov, V. S. "Problemy preobrazovaniia sel," Arkhitektura SSSR, 1971, no. 12.

Shkvarikov, V. "Problemy rasseleniia na sovremennom etape," Arkhitektura SSSR, 1970, no. 10.

Strogina, M. L. Sotsial'no-ekonomicheskie problemy razvitiia bol'shikh gorodov v SSSR, Moscow, 1970.

Urbanization in Socialist Countries

Svetlichnyi, B. "Razmyshlaia o sudbakh gorodov," Arkhitektura SSSR, 1967, no. 4.
Vladimirov, P., and Kochetkov, A. "Problemy perspektivnogo rasseleniia," Arkhitektura SSSR, 1974, no. 5.

Urbanization in Poland

Chmielewski, J., and Syrkus, S. Warszawa funkcjonalna, Warsaw, 1934.
Dangel, J. Przekstałcenia sieci miejskiej w Polsce pod wpływem rozwoju ludności i uprzemysłowienia kraju w okresie 1946-1960, Warsaw, 1968.
Dziewoński, K. "Hipoteza przeksztalceń sieci osadniczej w Polsce do roku 2000," in Polska 2000, Prognozy rozwoju sieci osadniczej, Warsaw, 1971.
_____. "Emerging Patterns of Urbanization in Poland," Geographia Polonica, 1972, no. 24.
_____. "Die Stellung der Ballungsgebiete im Siedlungssystem der Volksrepublik Polen," Petermanns Geographische Mitteilungen, 1973, no. 4.
Goldzamt, E. Urbanistyka krajów socjalistycznych, Warsaw, 1971.
_____. Modele przemian zespołów osadniczych, Warsaw, 1978.
Goryński, J. Supermiasto czy suburbanizacja, Warsaw, 1967.
_____. "Urbanizacja w punkcie wzrotnym," Gospodarka planowa, vol. XXXI, no. 12, 1976, pp. 663-65.
Grabowiecki, R. "Plan przestrzenny Polski do roku 1990," Gospodarka planowa, 1974, no. 4.
Jalowiecki, B. "Charakterystyka procesów urbanizacji Polski," Studia socjologczne, 1978, no. 3, pp. 101-26.
Karlowicz, R. "System wielkich aglomeracji miejskich w Polsce," Miasto, 1972, no. 12.
Kukliński, A., and Najgrakowski, M. "Industrializacja, urbanizacja i rozwój regionalny w Polsce," Miasto, vol. XXVIII, no. 1-2, 1979, pp. 4-16.
Leszczycki, S.; Eberhadt, P.; and Herman, S. "Prognoza przestrzennego zagospodarowania kraju do roku 2000," in Rozwój społeczny Polski w pracach prognostycznych, Warsaw, 1971.
Malisz, B. Zarys teorii ksztaltowania układów osadniczych, Warsaw, 1966.
_____. "Teoretyczny model sieci miast w Polsce," in Problemy rozwoju sieci osadniczej, Warsaw, 1967.
_____. "Prognóza zmian sieci osadniczej w Polsce," in Polska 2000. Prognozy rozwoju sieci osadniczej, Warsaw, 1971.
_____. Problematyka przestrzennego zagospodarowania kraju, Warsaw, 1974.
_____. Ewolucja struktury przestrzennej kraju, Warsaw, Omega, 1976.
Mrzygłod, T. Polityka rozmieszczenia przemyslu w Polsce 1946-1980, Warsaw, 1962.
_____. Zalozenia przestrzennego zagospodarowania kraju w okresie perspektiwicznym do 1985 r., Warsaw, 1968.
Planowanie przestrzenne — Plan krajowy I, Warsaw, 1947.

Selected Bibliography

Plan przestrzennego zagospodarowania kraju do roku 1990, Warsaw, 1974.
Stasiak, A. "Specyfika polskiej drogi urbanizacji w świetle wyników NSP," Miasto, 1971, no. 6.
Wallis, A. "Uwagi o społecznej problematyce Planu przestrzennego zagospodarowania kraju do r. 1990," in Procesy urbanizacji kraju w okresie XXX-lecia Polskiej Rzeczypospolitej Ludowej, J. Turowski, ed., Warsaw, 1978, pp. 67-78.
Zaremba, P. "Proba prognozy rozwoju sieci osadniczej w Polsce," in Polska 2000, Prognozy rozwoju sieci osadniczej, Warsaw, 1971.
_____. "Niektóre społeczne aspekty koncepcji urbanizacji Polski," in Procesy urbanizacji kraju w okresie XXX-lecia Polskiej Rzeczypospolitej Ludowej, J. Turowski, ed., Warsaw, 1978, pp. 51-56.
Zawadzki, St. M. "Zagospodarowanie przestrzenne Polski. Rok 2000. Założenia prognozy," in Polska 2000, Prognozy rozwoju sieci osadniczej, Warsaw, 1971.

Urbanization in the German Democratic Republic

Grimm, F. "The Settlement System of the German Democratic Republic. Report to the Commission on National Settlement Systems of the International Geographical Union" (mimeographed), Leipzig, 1979.
Grimm, F., and Hönsch, I. "Zur Typisierung der Zentrem in der DDR nach ihrer Umlandbedeutung," Petermanns Geographische Mitteilungen, vol. 118, no. 4, 1974, pp. 282-88.
Grimm, F.; Krönert, R.; and Lüdemann, H. "Aspects of Urbanization in the German Democratic Republic," in National Settlement Strategies East and West, Laxenburg, Austria, IIASA, 1975.
Grimm, F.; Lüdemann, H.; and Weinhold, P. "Selected Bibliography on Problems of Urbanization and the Development of Settlement Systems in the GDR (1970-1974)," in National Settlement Strategies East and West, Conference Proceedings, Laxenburg, Austria, IIASA, 1975, pp. 193-219.
Känel, A. V., and Scholz, D. "Wirtschaftsräumliche Struktureinheiten mittlerer Ordnung in der DDR," Petermanns Geographische Mitteilungen, 1969, vol. 3.
Kluge, K. "Die Bedeutung der Siedlungskategorien für die Planung der Siedlungsstruktur," Petermanns Geographische Mitteilungen, vol. 118, no. 4, 1974, pp. 255-60.
Krönert, R. "Functional Urban Regions in the German Democratic Republic," Laxenburg, Austria, IIASA, 1978.
Lüdemann, H., and Heinzmann, J. "The Interrelations between the Process of Concentration and Territorial Development under Socialist Conditions in the GDR," in Regional Studies, Methods and Analyses, Budapest, IGU, 1974.
Ostwald, W. "Planmässige proportionale Gestaltung der Territorialstruktur der DDR als Beitrag zur Erhöhung der Effektivität der gesellschaftlichen Reproduktion," Wirtschaftswissenschaft, 1976, no. 5.
Ostwald, W., and Scherf, K. "Die Siedlungsstruktur der DDR als ein wesentlicher Bestandteil der Territorialforschung und Territorialplanung," Petermanns Geographische Mitteilungen, 1974, no. 4.

Urbanization in Socialist Countries

Scherf, K. "Zu einigen ökonomischen, sozialen und territorialen Aspekten der sozialistischen Urbanisierung in der DDR," Wirtschaftswissenschaft, vol. XXVII, no. 1, 1979, pp. 31-46.

Scholz, D. "Die wirtschaftliche Struktur der DDR," Geographische Berichte, 1971, no. 2.

Sorgenicht, K., and Steglich, L. Gemeindeverbände – warum – wie – wozu?, Berlin, 1976.

Urbanization in Hungary

Beluszky, P. "Izmeneniia nastupivshie v strukture seti naselennykh tsentrov v Vengrii," in L. Dinev and N. Michev, eds., Problemy na geografiiata na naselenieto i selishchta, Sofie, 1973.

Bencze, I. "The Role of Capital Cities in Socio-economic Development," in Regional Studies, Methods and Analyses, Budapest, IGU, 1974.

Fodor, L. "Le développement des grandes agglomérations en rélation avec la croissance économique," in Urbanization in Europe, Budapest, IGU, 1975.

Fodor, L., and Illés, I. "Metropolitan Industrial Agglomeration," Regional Science Association Papers, 1968, vol. 22.

Köszegfalvi, G. "Einige Probleme der künftigen Entwicklung des Siedlungsnetzes in der Ungarischen Volksrepublik," Deutsche Architektur, 1964, no. 10.

Lettrich, E. Urbanizálódás Magyarországon, Budapest, 1965.

―――――. "Les traits caractéristiques de l'urbanisation en Hongrie," in Urbanization in Europe, Budapest, IGU, 1975.

Perczel, K. "A városközpontok országos rendszere és munkamegosztása," Területrendezés, 1972, no. 3.

Urbanization in Romania

Bogdan, T.; Cernea, M.; et al. Procesul de urbanizare in Romania. Zona Brasov, Bucharest, 1970.

Cănilă, C. "Aspecte ale urbanizării in R. S. România," Terra, 1974, no. 4.

Constantinescu, M. "The Urbanization Process in the Socialist Republic of Romania," Seventh World Congress of Sociology, Varna, 1970.

Cucu, V. "Romania," in Essays on World Urbanization, R. Jones, ed., London, 1975.

Deica, P., and Stefănescu, I. "Forms of the Territorial Grouping of the Settlement Network in the Socialist Republic of Romania," Revue roumaine de géologie, géophysique et géographie. Serie de géographie, 1972, no. 2.

Gusti, G. "Náměty na uspořádání struktury osídlení v Rumunsku" (translated in) Architektura, 1968, no. 4.

Rusenescu, C. "L'urbanisation et les nouveaux rapports entre la population, le site et le territoire de villages roumains," in Urbanization in Europe, Budapest, IGU, 1975.

Selected Bibliography

Sebestyén, G. "Některé předběžné předpoklady a hypotézy prognózy procesu urbanizace v RSR," translated in Architektura, 1972, no. 3-4.

Trebici, V. "Reflectii demografice in legatura cu prognoza populatici urbane," Architektura, 1972, no. 3-4.

Urbanization in Bulgaria

Grekov, P. "Saobrazheniia za tendenciite v razvitieto na gradoustroistvoto," Izvestiia na nauchnoizsledovatelskiia institut po gradoustroistvo i arkhitektura, vol. XXIII, no. 5, Sofia, 1971.

Michev, N. "L'urbanisation et le probleme du développement du réseau urbain en Bulgarie," in Urbanization in Europe, Budapest, IGU, 1975.

Problemi na geografiiata na naselenieto i selishchata, Sofia, 1973.

Tonev, L. "Funkcionalna organizacija i struktura na selishtnite edinici, grupi i sistemi," Arkhitektura (Sofia), 1973, no. 2.

Urbanization in Yugoslavia

Ginić, I. Dinamika i struktura gradskog stanovištva Jugoslavije, Belgrade, 1967.

Kokole, Vl. "Osnove policentričnega urbanega sistema v SR Sloveniji," Zavod SRS za regionalno prostorsko planiranje, no. 29, Ljubljana, 1975.

Krašovec, St. "The Future of Part-Time Farming," in Proceedings of the Twelfth International Conference of Agricultural Economists, Oxford, 1965.

Rančić, M. "Značenije i problemy demografičeskich prognozov naselenych punktov v Jugoslavii," Mezinárodní kongres demografů socialistických zemí, Liblice, 1976.

Štefanović, D., and Žuljić, S. Ekonomski aspekti izgradnje gradova u Jugoslavii, Belgrade, 1966.

Žuljić, S. Proces urbanizacije na prostoru Jugoslavije — Značenje i predvidivi tok promjena do 1985 godine, Zagreb, Ekonomski Institut Zagreb, 1970.

INDEX OF NAMES

Akhiezer, A. S., 12

Bauer, J., 18
Belousov, B., 66
Beluszky, P., 114
Blažek, M., 26
Bocharov, J., 62

Cănilă, C., 139
Čelechowski, G., 31
Chmielewski, J., 82-83, 84
Christaller, W., 119

Dangel, J., 75
Davidovich, V. G., 62
Deica, P., 129
Dziewoński, K., 84, 93, 97

Eberhardt, P., 76, 93
Engels, F., 57, 59
Enyedi, G., 126

Fomin, G. N., 71

Ginić, J., 141
Ginzburg, M. J., 59
Goldmann, J., 24
Goldzamt, E., 59, 61, 80, 154
Goryński, J., 84, 85
Grabowiecki, R., 11, 79, 92
Grekov, P., 139
Grimm, F., 110
Gusti, G., 136-37

Hampl, M., 10
Heinzmann, J., 108
Heřman, S., 76, 93

Honsch, I., 110
Hruška, E., 22
Hušek, P., 24

Ianitski, O. N., 12

Känel, von A., 104
Kaplan, G., 68
Karlowicz, R., 93, 94
Khorev, B. S., 51-52, 54, 72
Kniazev, K. F., 62
Kochetkov, A., 68
Kogan, L., 12
Kokole, V., 147
Kukliński, A., 4
Kumpošt, J., 22

Lakomý, Zd., 31
Le Corbusier, 59
Lenin, V. I., 57
Leszczycki, S., 76, 79, 84, 93, 94, 97
Lettrich, E., 114, 118
Líkař, O., 11
Listengurt, F. M., 68, 71
Lüdeman, H., 108

Malisz, B., 76, 81, 83, 84, 85, 87-88, 93, 97-99
Mareš, J., 18
Marx, K., 57, 59
Matoušek, V., 129
Michalec, J., 42
Mihailovič, K., 3, 142
Mohs, G., 111
Mrzygłód, T., 84, 85-87
Musil, J., 11

Author Index

Naimark, N. I., 71
Nowakowski, J., 84

Ostwald, W., 108

Pchelintsev, O. S., 72-73
Perczel, K., 123
Pióro, Z., 84
Pivovarov, Iu. L., 4, 10, 50
Probst, A. E., 62, 153

Riazanov, V. S., 63-64
Rybicki, P., 84

Scherf, K., 106
Schmidt, R., 111
Scholz, D., 104, 111
Sebestyen, G., 136, 139
Secomski, K., 84

Skvortsov, A., 62
Smidovich, S., 72
Stefănescu, J., 129
Střída, M., 23
Svetlichnyi, B., 64
Syrkus, Sz., 83

Taut, B., 59
Tonev, L., 138

Voženílek, J., 22

Weber, A., 57

Zaremba, P., 93, 98-99
Zawadski, S. M., 84, 92
Zhivkov, T., 137-38, 139
Źiólkowski, J., 11, 84
Zverev, R., 62

INDEX OF SUBJECTS

Accessibility of facilities, 151
Administrative reform in Poland, 98-99
Agglomeration effects, 8, 9
Agglomerations
 advantages and disadvantages, 108-9
 improvement of their environmental conditions, 151
 regulation of their growth, 151
 role in urbanization strategies, 157
Agglomerations in Czechoslovakia
 as basis for urbanization, 34
 categories, 34-36
 population projections, 36, 38-39
Agglomerations in GDR
 present state, 103-4
 modernization of, 108
 role in settlement strategy, 108
 role in the intensification of economy, 108-9
Agglomerations in Poland
 present state, 79
 role in settlement strategies, 92, 94
 spatial changes of, 94
Agglomerations in Romania, 129
Agglomerations in the USSR
 agglomeration processes, 49-50
 and the group settlement system, 69
Agglomerations in Bulgaria and their role in settlement strategy, 138
Agroindustrial settlements
 in areas of low density, 156
 in the USSR, 63-64
 in Yugoslavia, 146

Capital cities
 control of growth, 136, 139
 concentration of activities in Budapest, 118-19
 differences in regulation policies toward, 154-55
 function of Budapest in settlement strategy, 124
 growth of Sofia, 133-34
 regulation of Prague's growth, 26
 regulation of the growth of Budapest, 118
Central places
 categories of in GDR, 109-11
 categories of in Hungary, 124
 number of in GDR, 110-11
Central-place theory
 application in Czechoslovakia, 28-30
 application in GDR, 109-10
 application in Hungary, 123
 application in Slovenia, 147
 critique in Czechoslovakia, 31
Commuting in Czechoslovakia and GDR, 21
Comprehensive planning, 149
Concentration
 advantages of, 151
 as instrument of economic efficiency, 157
 aspects of, 10-11
 changing views on desirable levels, 157
 effects of, 10-11, 13
 excessive, 62
 of activities, 157
 of rural settlement, 151

Subject Index

"rational," 13, 124, 156
stimulation of, 43
Concentration processes
  in Bulgaria, 135
  in Czechoslovakia, limits, 33-34
  in GDR and agglomerations, 111-12
  in Hungarian urbanization strategies, 122-23, 125
  in Polish urbanization strategies, 85
  in the USSR, of rural settlements, 63
Concentration strategy
  based on agglomerations, 93-94
  moderate form in Poland, 87-88, 89, 91-92, 97, 99
Cooperating communities, 137, 156

Deagrarianization, 141-42
Decentralization of industry, 121
Decentralization of settlement
  and elimination of differences between town and country, 60
  radical, 59, 60, 153
  rejection of in the USSR, 60
  town planning ideas on, 60
Deconcentration strategy
  in Poland, 85
  in Slovenia, 147
Deglomeration policies in Poland, 86
Development axes, 139, 157

Economic and social goals balance, 149
Elimination of differences between town and country, 151, 155-56
Even distribution in settlement development, 152-53

Fragmentation of rural settlement, 63, 131, 136, 137

Group settlement system
  and division of labor, 68
  and integration of urban functions, 39
  and scientific-technological revolution, 69
  categories of, 70-71
  criticism of, 72-73
  definition of, 68

goals of, 67
impact on town planning, 72
in Romania and Bulgaria, 137-38
in rural areas, 9, 156
in the Ukraine and Urals, 71-72
official documents on, 66-67
Growth of cities, 45, 61-62, 101

Hierarchy in the settlement systems, 158
  efforts to avoid it, 122-23

Industrialization
  and urbanization, 6-7
  differences in, 138
  in Bulgaria, 132
  in Czechoslovakia, 18, 20-21
  in GDR, 100-101
  in Hungary, 114
  in Poland, 81, 83
  in Romania, 127
  in the USSR, 54, 58

Location of industry in Czechoslovakia, 18, 20-21, 23, 24, 26
Location policies
  as basis of the settlement strategy in GDR, 105
  common features in socialist countries, 149
  in Bulgaria, 138
  in the Party Congress decisions in the USSR, 58-59
  in the resolutions of SUP in GDR, 105
  rational distribution of production, 151
  under socialism, 56-58

Medium-sized towns
  stimulation of their growth, 139
  their growth in Yugoslavia, 144
Migration into cities
  and "ruralization" of cities, 7
  in Bulgaria, 87, 97-99, 111, 139, 157
  in Czechoslovakia, 75-76
  in Poland, 75-76
  in Romania, 127-28

Nodal belt system of settlement, 87, 97-98, 111, 139, 157

Subject Index

Nodes
  role in future settlement, 94-97
Optimum size of cities, 62-63

Peasant workers, 7, 82, 142
Physical planning legislation
  in Czechoslovakia, 31-32
  in Poland, 83-84
Physical planning practice
  relation to economic and sectoral planning, 15, 27-28, 66, 83, 88
Polarization theory, 14
Polycentric concentration, 11, 58, 89-92
  as decentralized concentration, 125-26, 139, 147, 157
Population growth and settlement in Hungary, 114, 115-116
Postcity settlement system, 73

Regional development
  in Czechoslovakia, 25, 26
Regional differences
  different approaches to, 153-54
  general aspects of, 12-13
  in Czechoslovakia, 18, 20, 23-24
  in GDR, 104, 107-8
  in Hungary, 118, 119-20
  in Poland, 79, 81
  in Romania, 132
  measures to eliminate them, 26-27, 39
  reduction of them, 149, 152-53
Regional prognoses, 71
Ruralization of cities, 7
Rural population
  decline in Bulgaria, 132-33
  decline in Yugoslavia, 142
  growth, in Poland, 75
Rural settlement system
  depopulation of, 112
  future in GDR, 111-12
  in Romania and Bulgaria, 136
  on the Great Hungarian Plain, 118-19

Self-management system and settlement, 142, 144
Settlement agglomerations in Czechoslovakia
  and urbanization, 34, 39
  lower-level categories of, 35
  regional, 34-35
  types of, 34
Settlement strategies
  in Czechoslovakia, 27, 32
  in Hungary, 119-21, 122
  in Poland, 84-85, 87-89
  in Romania and Bulgaria, 136-37
  in Yugoslavia, 146-47
  permanent and common features in socialist countries, 148-51
  why they differ, 148
Settlement structures
  differences in Poland, 79-80
  economic effects, 24, 156
  in newly opened territories, 64-66
Settlement system
  and production, 30
  and tertiary sector, 25
  dynamic character of, 158
  in Czechoslovakia, 17-18
  in GDR, 100-101
  in Hungary, 113, 114, 116, 118-20, 124
  in Romania, 131-32
  in the USSR, 53-56
  in Yugoslavia, 144-46
  theory of in GDR, 105
Settlement zones
  and group settlement system, 69-70
  in the GDR, 103, 106-7
  their activization in the USSR, 155
  their future role in Poland, 98
  their problems in Yugoslavia, 144
Strong regional centers, 151

Territorial rationalization, 108

Urban and rural sociology, 7-8
Urban regions
  their role in settlement strategy, 97
Urbanization
  and agriculture, 9, 117-18
  and industrialization, 6-7, 100-101, 151-52
  and modernization, 7
  and scientific technological revolution, 6

Subject Index

areas, old and new, 82
as condition for intensive economic development, 106
balance of effects, 157
definition of, 5
indirect, 151, 156
limiting factors, 158
managed, 6, 104
metropolitan
  in Bulgaria, 133, 135
  in Hungary, 113
  in Romania, 128-29
of the countryside in Poland and Yugoslavia, 32, 144-46
specific forms of, 6
Urbanization in Bulgaria
  high rates of, 131-32
  trends up to 2000, 139-40
Urbanization in Czechoslovakia
  level of, 17-18
  postwar trends, 20-21
  the trends up to 2000, 36-39, 44
Urbanization in GDR
  level of, 103
  main future trends, 111-12
Urbanization in Hungary
  level of, 116
  trends up to 2000, 125-26
Urbanization in Poland
  as a historical process, 93-94
  estimated level of by 2000, 87
  level of in the postwar period, 75, 76, 77
Urbanization in Romania
  level of, 127
  rate of, 127
  trends up to 2000, 139-40
Urbanization in the USSR
  high rates, 45-49
  history and stages of, 46-49
  in different settlement zones, 51-53, 54
Urbanization in Yugoslavia
  level of, 141
  of individual republics, 144-45
Urbanization policies
  differences of, 152
Urbanization strategies
  for agricultural regions in GDR, 109
  main objectives in the USSR, 71
  main features in Slovakia, 40-44
  postwar in Poland, 81

ABOUT THE AUTHOR

Born in 1928 in Ostrava, Czechoslovakia, Jiří Musil was educated at the Charles University, from which he received his doctorate in sociology in 1952. From 1952 to 1958 he did research for the Czech Institute of Public Health and then moved to the Institute of Building and Architecture, where he is now head of the Department of Sociology. A frequent visitor to Western universities, Jiří Musil was a research fellow at the University of Glasgow and the London School of Economics, has taught at the Hannover University, has lectured at the University of Kent, and has been a consultant for the Housing Committee of the Economic Commission for Europe in Geneva.

Dr. Musil is the author of numerous books and articles on various aspects of housing, urban planning, and the problems of cities.